A FISHKEEPER'S GUIDE TO

SOUTH AMERICAN CICHLIDS

A splendid survey of this attractive and diverse group
of freshwater tropical fishes

Dr Wayne S Leibel

A Salamander Book

Published in North America by TETRA PRESS
3001 Commerce Street
Blacksburg, VA 24060

ISBN 1-56465-103-7

All correspondence concerning this volume
should be addressed to TETRA PRESS.
WL # 16013

Printed in Belgium
First edition 1993

639.344

Credits

Editor: Tony Hall
Designer: Paul Johnson
Color artwork: Gordon Munro ©Salamander Books Ltd
Color reproductions: P&W Graphics PTE Ltd, Singapore
Filmset: The Old Mill
Printed in Belgium by Proost International Book Production

Author

Dr Wayne S. Leibel has been an aquarist for over 35 years, specializing in fish of South and Central America for the past 15 years. He took his Master's and Doctorate Degrees from Yale University in comparative biochemistry and was a Research Fellow at Harvard University before taking up his present position as an Associate Professor of Biology at Lafayette College in Pennsylvania (USA). His current research concerns the molecular evolution of Neotropical cichlid. Dr Leibel has served two terms on the Board of the American Cichlid Association and for seven years edited their Journal *The Buntbarsche Bulletin*. He was recently made a Fellow of the ACA for his efforts. He has collected fish in both, South and Central America, and has spawned, photographed and written about many of the rare cichlids he has encountered. Despite his recent professional involvement with fish, he remains a hobbyist at heart.

Contents

Introduction

Out of the huge variety of tropical fish available today, cichlids have captured the hearts and interest of a steadily growing number of aquarists because they combine aesthetics with interesting behaviour. While true of cichlids in general, this seems especially so with cichlids of the New World due to their highly-developed pair-bonding and parental behaviours. The majority of South and Central American cichlids form stable monogamous pair bonds through elaborate physical and visual behavioural rituals, and exercise diligent, even gentle, care of their eggs and offspring. These courtship and spawning behaviours are a joy to behold and constitute one of the main attractions of keeping Neotropical cichlids in the home aquarium.

The 100-plus cichlid species of Central America have radiated fairly recently from a common ancestral Cichlasomine that presumably emigrated from the south, and thus are quite similar in appearance and habits. In contrast, the more ancient South American cichlid fauna is distinctive for its variety: with several million years' more time for evolution to have wrought its magic, cichlid fishes have adapted to exploit a wide variety of ecological niches via dramatic changes in body shape and in reproductive and feeding behaviours. Thus have evolved the highly-compressed Angelfish and Discus, the torpedo-shaped piscivorous pike cichlids, the bizarre Geophagine Eartheaters that sift the bottom for food and practise various forms of mouthbrooding, the

cave-dwelling dwarf cichlids, the bottom-hopping rapids dwellers, alongside the more typical Cichlasomine species reminiscent of and directly related to those of Central America. This variety spells sustained interest for the budding cichlidophile who must face and solve varying problems peculiar to particular species and their specific life histories. There is no single formula for maintaining and spawning South American cichlids, but in that lies their charm and the challenge.

In this book we provide both general information concerning the maintenance of generic South American cichlids and particular strategies for various species groups. The more general information presented in the first part of this guide will be amplified on a species-by-species level in the second part. Remarks in the latter will, by necessity, be confined to only 40 species, a fraction of the total South American cichlid species currently available today in the aquarium hobby. However, they represent a starting point from which you can proceed with other members of the various groups.

Part of the joy of keeping tropical fish, and cichlids in particular, is discovering the requirements of individual species and creating the conditions conducive to their captive breeding. Approached in this way, the aquarium hobby can become a lifelong appreciation of the beauty and mystery of nature. The cichlids of South America certainly provide ample grist for the discovery mill.

South America and its cichlids

South America as we know it today is half the size of Africa and one-third the size of Eurasia, being some 7250km (4500 miles) long and 4800km (3000 miles) across at its widest point — a total area of approximately 18 million square km (7 million square miles). Its geography is dominated by the world's largest rainforest which is drained by the second longest river in the world, the Amazon. The Amazon River delivers one-fifth of the total riverine freshwater discharge of this planet, the largest volume of water discharge and 7-10 times that of the Mississippi River. It drains two-fifths of South America, including Peru, Ecuador, Colombia, Venezuela, the Guianas, and Brazil. It boasts over 1000 tributaries, 17 of them over 1600km (1000 miles) long. In places its depth reaches 90m (295 feet), allowing ocean-going ships access to the Peruvian port of Iquitos, some 3700km (2300 miles) inland at the foot of the Andes, and it is as wide as 11km (7 miles) at some points.

Goulding (1990) estimates that a total 2500-3000 species of fish, only half of which have yet been formally described by ichthyologists, reside in the Amazon drainage. With a world total of 6650 freshwater fish, the Amazon basin is the world's richest ichthyofaunal region with nearly 10 times as many fish

species as all of Europe. Of that total, Lowe-McConnel (1991) suggests that nearly 225 species, 75 per cent of the total estimated 300 species of South American cichlids, are found in the Amazon drainage. However, cichlids make up only about 6-10 per cent of the total fish diversity of the Amazon, with characoids (tetras, silver dollars, etc: 43 per cent) and catfishes (Siluriformes: 39 per cent) being the most highly-radiated and represented groups.

South American climate and geography

Most South American cichlids are lentic fishes, inhabitants of slow-moving or stagnant waters, and, with certain exceptions, they tend not to frequent the rapidly-flowing channels of rivers. Rather, they are found in smaller water bodies — tributaries, streams, creeks, or associated pools, oxbow lakes, laguna, marshes and the like. They get there during the rainy season, which lasts from November to June, when the Amazon rises as much as 15m (50 feet) and spills over its banks to flood adjacent forest. The flooding extends as much as 80-95km (50-60 miles) beyond the normal river channel and encompasses an area of some 100 square km (38,600 square miles) — about 2 per cent of the total Amazon rainforest (*igapo*) and an area larger than England. This flooding happens because the Amazon basin is so flat, rising no

Below: *During highwater, the rivers spill over their banks to flood the forests.*

more than 200m (650 feet) above sea level at its highest point. Thus, the massive volume of water caused by the torrential rains and melting ice in the Andes Mountains flows eastward towards the Atlantic Ocean and spills over the river banks that are unable to contain the swell. This seasonal flooding, more than anything else, dictates the rhythm of life in Amazonia. The flooded forests are extremely important spawning grounds for most Amazonian fish, including cichlids, providing food in the form of insects and fallen fruits, and cover for newly-hatched fry. After the peak of the flooding, some fish simply escape the draining forest by returning to the river channel. Others remain in marshes or *varzea* lakes during the subsequent dry season. These seasonal floods cover vast expanses of Amazonia for 4-7 months each year.

Although it is central, the Amazon Basin is not the entire story of South America and its fishes. The Amazon Basin is bounded by the Andes Mountains to the west, the Brazilian Highlands to the south and east, and the Guiana Highlands to the north. To the west of the Andes, great deserts, 2570km (1600 miles) of them, extend from the mountains down to the sea. The Brazilian Highlands are formed of ancient crystalline rock and cover 1,500,000 square km (580,000 square miles) of primarily arid savannah consisting of scrub forest and grasslands, of which the Mato Grosso is the most famous. As the plateaux diminish in altitude towards the south, the tropical deciduous forest and scrub of the Gran Chaco dominate the landscapes of Bolivia, Paraguay and northern Argentina. The partly swampy plains or *chaco* grade southward into the Argentine *pampas*, or vast arid grassland regions. To the south, along the eastern base of the Argentine Andes, is a large desert with shrubs and cacti. To the south and west, in the shadow of the Chilean Andes, humid temperate (subtropical) forests characterise the

Above: *During the rainy season, the rivers rise as much as 15 metres (50 feet). The mud coating on this tree denotes the highwater mark.*

region known as Patagonia which, as you travel southward, increasingly resembles Alaska, complete with glaciers. Southernmost South America, Tierra del Fuego, is decidedly cold and humid, and is the winter home to several species of penguins.

Clearly South America is a continent of extreme contrasts: of tropical rainforests which occupy nearly half of the landmass, but also of mountains and deserts, of huge grassland savannahs and subtropical deciduous forests. While the average yearly temperature of Amazonia is 27°C (81°F) (with extremes of 20-37°C (69.4-98.4°F) recorded at Manaus, Brazil), Uruguay to the south is decidedly temperate, averaging 10°C (50°F) in winter and 22°C (71°F) in summer. Average yearly temperature variations from 0-9°C (32-49°F) have been reported from Tierra del Fuego at the tip of the continent. Not surprisingly, South America consists of a range of 'life zones' supporting a wide variety of organisms adapted to the highly variable topography and associated climate. Since rivers penetrate most of the South American continent, it should come as no surprise that the fishes of South America are equally diverse and adapted to these zones.

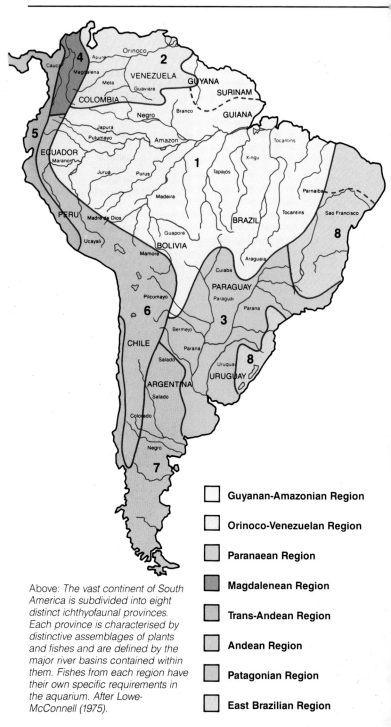

Above: *The vast continent of South America is subdivided into eight distinct ichthyofaunal provinces. Each province is characterised by distinctive assemblages of plants and fishes and are defined by the major river basins contained within them. Fishes from each region have their own specific requirements in the aquarium. After Lowe-McConnell (1975).*

Legend:
- Guyanan-Amazonian Region
- Orinoco-Venezuelan Region
- Paranaean Region
- Magdalenean Region
- Trans-Andean Region
- Andean Region
- Patagonian Region
- East Brazilian Region

The ichthyofaunal regions

The French ichthyologist Jacques Gery recognised eight major ichthyofaunal regions in South America. The Amazon drainage is more correctly known as the Guyanan-Amazonian region because the Guianas (Guyana, French Guiana, Surinam) and the Amazon basin share fish across the seasonally-flooded savannahs of Guyana to the north. Similarly, the second major region, the Orinoco-Venezuelan region to the northwest, is also connected to the Amazon system via the Casiquiare Canal (Rio Negro, Brazil; Rio Orinoco, Venezuela) during highwater. The Rio Orinoco, which originates in Colombia and travels north through Venezuela to the Caribbean Sea, flows through the flat, often marshy grasslands, or *llanos*, which flood during the rainy season and exchange fish with the Rio Negro. Nearly 90 per cent of the South American cichlid fauna is distributed between the two regions.

The third major ichthyofaunal region is the Paranaean. Cichlids are also found here as well, though not with the richness found in Amazonia. It is the second largest drainage in South America and comprises the La Plata-Rio Uruguai-Rio Parana-Rio Paraguai system in Paraguay, Uruguay and Argentina. In fact, the confluence of these rivers at La Plata makes this compound 'river' (Rio de la Plata) the second largest in the world, after the Amazon, in terms of total water volume discharge. The Paranaean region is very dry and seasonally swampy and vast marshes known as the Pantanal flood annually. Although isolated today, it was connected with the Amazon, so many of the fish species are now widely distributed from the Rio Orinoco down to this essentially temperate zone. Not surprisingly, these species have become 'coldwater' adapted and are especially tolerant of lower temperatures in the aquarium.

To the north and east of the Paranaean region, and forming the eastern boundary of Amazonia (the Brazilian Highlands), is the East Brazilian ichthyofaunal region, characterised by smaller rivers that flow eastward to the Atlantic coast. While the immediate coast is humid and covered by dense forest, the highlands themselves are arid plateaux which are largely scrub forest or even desert. Many of the coastally-distributed cichlids venture into partially-saltwater estuarine habitats and can tolerate, even require, harder waters in contrast to the generality of softwater Amazonian cichlids.

The four remaining regions are not particularly rich in fish — or cichlid — species. The Magdalenean and Trans-Andean regions of northwestern South America, Colombia and Ecuador are defined by the Andes which split into three distinct massifs in Colombia. The Magdalenean is east of these mountains and the Trans-Andean is west. There are some, but not many, cichlids found here. The Andean and Patagonian regions are nearly completely devoid of fishes due to the exceedingly low temperatures. However, one of the pike cichlids, *Crenicichla lacustris*, is sometimes found in the 'warmer' parts of Patagonia.

Understanding biotopes

From an understanding of the basic biogeography of South America and its cichlid fauna comes a strategy for maintaining these fish successfully in the aquarium. Cichlids have been quite successful in exploiting most environmental niches in both tropical and some subtropical waters of this continent. They have adapted to a variety of lifestyles, from sifting benthic invertebrates (Eartheaters) to ambushing other living fishes (Pike Cichlids, Basketmouths) to eating fruits that fall into the water from above (some Cichlasomines) and everything in between. An appreciation of the geography and biotopes where cichlids live is exceedingly helpful in guiding the aquarist's efforts.

South American cichlids in the aquarium

Brushing up on the basics
There are a few basic rules with which all aquarists, regardless of the fish they keep, should be familiar. Before discussing the particular requirements of South American cichlids, it is useful to review these basics. We will do so here, but you are encouraged to seek further information in any of the excellent general hobby books.

Water chemistry
There are two major parameters of water chemistry of import to aquarists: pH and hardness. The pH of the water is a measure of its relative acidity or alkalinity, or more correctly, the concentration of dissolved hydrogen ions (H^+). The greater the concentration of hydrogen ions, the more acid is the water. The pH scale ranges from 0 (extremely acidic) to 14 (extremely alkaline). The pH of pure, distilled water is 7 which is said to be neutral. Because the scale is logarithmic, a decrease of the pH from 7 to a value of 6 is actually a ten-fold increase in the concentration of hydrogen ions. Water pH can be tested using simple colorimetric kits or portable electronic pH meters.

Freshwater fish typically live at a pH level of between 6 and 9, but some South American blackwater species require even more acidic water (pH 4-5). It is possible to adjust the pH of your water by using chemicals, but such changes should be done very gradually, if at all. Most advanced aquarists avoid changing pH and instead select fishes to suit their own water chemistry.

Water hardness refers to the amount of dissolved salts present in the water. The major salts that effect hardness include calcium, magnesium, carbonate and bicarbonate. Two types of hardness are measured and discussed: general or permanent hardness (GH) which measures calcium/magnesium levels, and carbonate or temporary hardness (KH) which measures carbonate/bicarbonate levels. Carbonate hardness is said to be temporary because it can be boiled off or removed by filtration through peat.

In Europe, water hardness is usually expressed in degrees hardness or dH. One degree of dH is equivalent to 10 mg of calcium oxide (CaO) or magnesium oxide (MgO) dissolved in one litre of water. In the USA, the situation is different and hardness is expressed in ppm (parts per million) of total dissolved solids and is equivalent to 1 mg of calcium carbonate ($CaCO_3$) dissolved in one litre of water.

Water is classified according to its dH as follows:

dH	$CaCO_3$	Hardness
0-3 dH	0-50 ppm	Soft
3-6	50-100	Moderately Soft
6-12	100-200	Slightly Hard
12-18	200-300	Moderately Hard
18-25	300-450	Hard
>25	>450	Very Hard

Water hardness may be measured using commercial chemical 'drop' lather tests, or with an electronic conductivity meter. The conductivity meter measures total dissolved salts and expresses them in microSiemens/cm: the more dissolved salts, the better electrical conductor the water becomes. Although conversion to dH is inaccurate, for aquarium purposes 35 microSiemens is equivalent to 1°dH. Water hardness may be altered to fit the needs of the fish being maintained.

Water quality
Perhaps more important than water chemistry is water quality — how 'clean' the water is kept. Nitrogenous wastes (ammonia, nitrites, nitrates) which are toxic to fish, build up rapidly in closed systems like the aquarium. In closed systems, methods must be found to remove these toxic compounds which are continually accumulating. Filtration and water changes provide the solution.

Filtration

There are several approaches to filtering water in the aquarium. Mechanical filtration is the sieving of suspended coarse waste particles — faeces, uneaten food, detritus — from the water and is achieved by passing aquarium water through mechanical filtration media, typically polyester filter floss. While mechanical filtration removes particles from the water, unless the floss is changed regularly, the decaying detritus will remain in the closed system and release toxins.

One way to remove dissolved toxins is through the use of chemical filtering materials like activated carbon or exchange resins that adsorb (bind) dissolved organics. However, activated carbons, which vary in quality and porosity, must be changed frequently because they tend to clog and can actually release toxins back into the aquarium when overloaded. Clay-like zeolites of the sort also used as cat litter and sold specifically as aquatic ammonia absorber may help remove this toxin. More recent high-tech bonded resins provide a still higher degree of non-specific chemical filtration. All non-selectively remove organics, including medications, so their use should be discontinued when medicating.

A second, more natural way of removing toxic nitrogen wastes is via biological filtration. In nature, bacteria participate in the normal breakdown of ammonia. Nitrifying bacteria (such as *Nitrosomonas* sp.) convert ammonia (NH_3) to the less toxic intermediate nitrite (NO_2-) which is in turn converted by bacteria such as *Nitrobacter* sp. to even less harmful nitrates (NO_3-). These nitrifying bacteria are aerobic. They exist naturally in the aquatic environment, but can be induced to colonise porous surfaces — gravel, carbon, plastic foam — which have a good flow of oxygenated water. This is the strategy of biological filtration: to encourage the proliferation of these 'good' bacteria by providing surfaces for their colonisation.

Biological filtration may be achieved through a variety of technical approaches. In undergravel filtration the gravel bed itself, raised off the bottom with plastic grating to allow free passage of water through the gravel, serves as the site of bacterial colonisation. Water is drawn down and through

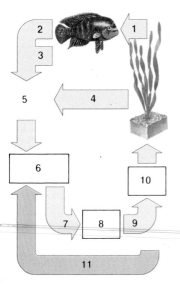

Left: The Nitrogen cycle. In nature, bacteria participate in the normal breakdown of nitrogen wastes. Nitrifying bacteria convert ammonia into the less toxic intermediate nitrite which is in turn converted by other bacteria to even less harmful nitrates. The nitrates can be used by plants. These "good" bacteria are aerobic.

Food (1) is eaten by the fish. Fish waste and uneaten food (2,3), together with plant waste (4) decompose under the action of fungi and bacteria. This action produces toxic ammonia (6). Under the action of aerobic nitrifying bacteria (7), the ammonia is converted into nitrites (8), which are in turn converted by other bacteria (9) into nitrates (10), which are used by the plant. The process is aided by some anaerobic bacteria (11).

the gravel bed for return via discharge tubes at the surface. Undergravel filters are not recommended for most Neotropical cichlids. A second, more effective approach is 'sponge' filtration which utilizes synthetic (plastic) foam as the colonising substrate. Air-lift or powerhead sponge filters are placed into the tank. Regardless of specific technology, biological filtration systems take several weeks to 'cycle', ie. for the proper bacteria to establish themselves in the porous media. To hasten the cycling, systems can be seeded with bacteria from an established tank by simply squeezing out the contents of working sponges into the new tank, or by using commercially-available seeding bacteria.

Two high-tech solutions have revolutionised fishkeeping: the canister filter and the wet/dry trickle filter. The canister filter is an external power filter whose filtration media are contained in a cylindrical sealed canister. The media include floss (mechanical) ceramic rings or foam block. A pump pulls water from the tank through the canister and returns it via a spray bar.

Wet/dry trickle filters, originally designed for marine tanks, have proven effective in freshwater tanks. These are also external, power-driven filters. The filter consists of a large acrylic box into which are stacked porous plastic balls, cubes, etc, which provide the colonisation bed for the nitrifying bacteria. These plastic media have a tremendous surface area and lots of hollow spaces that enable water to 'trickle' through them. The water from the tank, mechanically pre-filtered, is delivered via a rotating spray bar or a perforated drip plate to the stacked media and allowed to trickle through it (the 'dry' phase). In this aerobic environment, ammonia is detoxified and the water oxygenated. The trickled water collects in a second compartment, often passing through a foam filter, and is pumped back to the tank via a submersible pump.

There is no one filtration approach that is perfect. In fact, most successful aquarists use several filter types on each large tank. For instance, simple box filters providing mechanical filtration and aeration coupled with external power, or canister filters providing chemical/biological filtration.

Water changes

All filtration *must* be supplemented with regular partial water changes in the cichlid aquarium. As water evaporates, minerals and waste accumulate. If one simply 'tops off' the tank with other than pure, distilled water, the dissolved minerals will concentrate. Moreover, biological filters cannot usually handle all of the nitrogenous waste, particularly for larger cichlids. The

A thicket of plants provides hiding places favoured by these shy fish. Java ferns and floating plants are excellent choices.

Above: *A typical set-up for Dwarf Cichlids or medium sized acaras includes both live plants, spawning caves and bogwood shelter.*

solution is weekly or biweekly 20-40 per cent water changes. The water is simply siphoned off and replaced with tap water, the temperature of which is plus or minus a few degrees of the tank.

Before using raw water for water changes, however, it is a good idea to check its chemistry. In these days of acid rain, metropolitan water sources frequently 'lime' the water near the pumping source making it more alkaline so delivery pipes will not be harmed by its natural acidity. What comes from your tap may measure pH 7-9, but if it is very soft, it will have no buffering capacity and the pH will drop, harming your fish. You may have to draw water first into a reservoir and allow it to stand, or filter it with a biological (eg sponge) filter before use.

Pieces of cured bogwood provide shelter for these shy fish as well as a visual counterpiece for the aquarium tank.

Tank selection

Most beginners select smaller tanks, typically 45 litres (ten gallons), to start. That is a mistake. They should purchase the biggest tank they can afford and have space for. The reason is simple: a bigger volume of water gives a greater cushion in terms of managing water quality and catching problems before they become critical. For the Neotropical cichlid aficionado, there is another reason for selecting a larger tank. Most cichlid species are territorial and aggressive and need room if several are kept together. Moreover, tanks with larger bottom areas are preferable to 'high' (deep) tanks of the same volume for the same reason. In selecting cichlids for the tank, always take into account the

The use of undergravel filtration should be limited to small cichlids like apistos, which produce few wastes and do not dig.

Tipped flowerpots, or upended with a chink in the rim, provide spawning caves and shelter.

Fine gravel or sand is the substrate of choice for these small, non-digging cichlids.

eventual size of the fish, or plan to move them to larger quarters as they grow.

Substrate

Some thought must also be given to the selection of substrate. For dwarf varieties which require planted tanks and do not dig, fine gravel or even sand is appropriate: a substrate that will anchor rooted plants but will not pack unduly. This same substrate is also appropriate for the sand-sifting Eartheaters (*Geophagus* species). For larger cichlids, larger sized gravel is ideal. However, pebble-sized gravel packs too loosely and allows uneaten food and wastes to percolate into the spaces below the surface where they rot. For large, gravel-excavating species, bare tanks may prove the best, most hygienic, though less aesthetic, solution.

Shelter

Since cichlids tend to be aggressive and sometimes downright belligerent, it is appropriate to provide shelter where harassed individuals can escape. This can be a simple tangle of driftwood, a pile of rocks, broken or inverted clay flowerpots, plastic PVC piping of appropriate diameter and length, or any number of innovative and

Below: *Java Fern*, Microsorium pteropus, *is well suited for all but the rowdiest of cichlids.*

non-toxic solutions. Keep watch for signs of battering and be prepared to remove or separate combatants. In the wild, beaten fish simply swim away. In the confines of the aquarium, they are often killed.

Plants

Although many Neotropical cichlids have deserved reputations as diggers or plant-eaters, some should not. Most dwarf cichlids and many of the medium-sized Acaras actually prefer planted tanks. But even some of the 'diggers', do best in planted tanks. The ideal cichlid plant is one that provides the cover of rooted plants without requiring rooting, that tastes unpleasant to them, and which can take the higher temperatures of 26-29°C (78-84°F) that Neotropical cichlids seem to enjoy. Perhaps the ideal aquarium plant is the Java Fern, *Microsorium pteropus*. This wonderful deep green plant, with 20cm (8in) bladed leaves grows from a rhizome and has rhizoids rather than roots. It need not be rooted, but loosely attached to submerged bogwood. It propagates both by creating new plants along the margins of its leaves (offsets) and by rhizome extension. Luckily, cichlids seem not to enjoy eating this plant, but larger species will rip it to shreds. It is ideal for all but the largest, inveterate plant-destroying cichlids.
  Other useful and relatively easy

Left: Some fish, like this pike cichlid, require live 'feeder fish'. The wise aquarist will research his intended fishes' dietary requirements before purchase.

aquatic plants include Java Moss (*Vesicularia dubyana*), which also anchors to bogwood, and that old standby, Water Sprite (*Ceratopteris thalictoides*) which, as a floating plant, will cut down on top illumination. Duckweed (*Lemna* sp.) and Salvinia (*Salvinia* sp.) are also good top plants. Rooted aquarium plants, such as Amazon Swords (*Echinodorus* sp.) may be potted up with large pebbles or stones as the top layer to discourage uprooting. Plastic plants are another alternative.

Lighting

Lighting is best provided by fluorescent bulbs. These are considerably cooler to run than incandescent bulbs and make temperature control less of a problem. In general, cichlids like moderate light intensity. Selection of bulb type (cool vs. warm white, grolux, full-spectrum) depends on aesthetics, budget, and whether the intent is to cultivate plants along with fish.

Foods and feeding

There is a long list of nutritionally-valuable foods available to the modern aquarist. These include prepared dried flake and pelleted foods, freeze-dried foods, frozen foods and, of course, live foods. The strategy in selecting appropriate foods requires knowledge of what and how particular fish eat in the wild. Some species are not finicky and, after a short period of time, will adapt to eating prepared foods. Other species resist all but live foods. Know your fish before buying them, and resist obtaining any species you will not be able to feed properly. In general, fish health depends on your ability to get them eating a *variety* of foods. Food is not a subject on which to be penny-wise and pound-foolish.

Frozen foods provide a reasonable compromise between the nutritive value of natural foods and the convenience of prepared foods. These days, a wide variety of collected, washed and flash-frozen invertebrates, including insect larvae (eg bloodworms, glassworms, mosquito larvae) and various crustacea (eg daphnia, brine shrimp, plankton, krill), are available in frozen blocks that are easily stored in the home freezer, thawed and fed to your fish. Freeze-dried versions of most of these are also available. In addition, a number of frozen 'paste' foods based on beefheart or liver, or various shrimp/fish combinations are also available.

Obviously, live foods are nutritionally the best foods. However, there is considerable work involved in the collection or culturing of these foods. Certain worms, including whiteworms (*Entrychaeus*) and earthworms (*Lumbricus*), can be raised in quantitites that make regular feedings practicable. Daphnia and Brine Shrimp (*Artemia salina*) can also be cultured, but with some difficulty. Live foods are often available from your dealer. These include Daphnia, adult *Artemia*, Glassworms and Tubificid (*Tubifex*, black) worms. Although the latter are particularly relished, they are collected from sewage and unless purged thoroughly in clean water before feeding, they can wreak havoc. Similar problems often accrue from feeding live 'feeder fish' (eg guppies, minnows, goldfish) to large cichlids. Sometimes the feeder fish themselves carry diseases that prove harmful or fatal to the fish that eat them. It is advisable to wean your fish off live feeder fish if you can.

Some South American cichlids, like Uaru or Severum, are partially herbivorous. For cichlids requiring plant material in their diet there are several options. The addition of spirulina or algae-containing prepared foods will provide some of the needed nutrition. However, a better solution is to offer romaine lettuce, spinach (either fresh or par-boiled) or sliced, par-boiled courgettes (zucchini squash). After getting used to these foods, herbivorous cichlids will eat them ravenously.

Disease
In general, Neotropical cichlids are hardy. Like most fish, they are susceptible to the usual list of common diseases like 'ich' (*Ichthyophthirius multifiliis*), flukes or finrot which respond to the usual treatments. However, there are a few diseases that seem peculiar to Neotropical cichlids. The hobbyist is often dealing with wild-caught

Below: 'Head Hole' or neuromast pitting in cichlids is often caused by lax water maintenance.

specimens. These often come in 'rough' and in need of some care. Often they are emaciated and pick at, but reject, all offered foods, live or otherwise, and produce white stringy faeces. This syndrome may result from any of the following organisms: the protozoans *Hexamita* or *Spironucleus*, or threadworms (*Capillaria*), or others. Metronidazole is the drug usually recommended for the treatment of flagellate protozoans in cichlids and Dylox (Masoten) for threadworms. Naladixic Acid, an antibiotic has also proven exceedingly useful. Finally, wild fish are often parasitised with gut worms. There are a number of medicated prepared foods on the market that pack several antihelminthics in palatable form which may be useful in 'worming' fish.

Another typically cichlid disease is 'Head Hole'. In this syndrome sensory organs of the head and lateral line called neuromasts erode or 'pit out'. Although the protozoans mentioned above were once believed to cause 'Head Hole', it seems that the problem is primarily environmental. Certain species are particularly sensitive to lax water quality maintenance and respond in this way. Sometimes the erosion is reversible with the resumption of regular water changes, but sometimes not. Moreover, even in the best of water conditions, some species will nevertheless 'pit out',

suggesting that something lacking in the water is triggering the condition.

Cichlids will fight and often the aquarist is faced with treating a variety of physical injuries. These include split fins, raised scales and, at worst, open wounds. Small injuries heal by themselves if the loser is isolated. However, open wounds quickly become a breeding zone for bacteria. Wounds may be treated by simply netting the fish and carefully swabbing the area with commercial merthiolate solution (mercurichrome). Allow the solution to soak in but avoid the gills. Repeat daily until the wound shows signs of healing. Mercurichrome should not be added to the water.

Finally, some species, in particular Eartheaters, seem prone to a syndrome called 'Neotropical Bloat' in which the affected fish will suddenly develop pronounced abdominal swelling, begin ventilating heavily, stop eating and die after hanging listlessly at the top of the tank for several days. Lax water maintenance seems not to be the cause nor is it contagious. Nothing seems to cure it.

Creating a South American habitat in your tank

The aim of any aquarist is to simulate a fish's normal habitat to ensure its health and well-being, often with captive propagation as

Above: The 'Wedding of the Waters' occurs where blackwater tributaries run into the main silt-laden whitewater channel of the Amazon.

the ultimate goal. While some of the commercially available South American cichlids are tank-reared and thus adaptable to a variety of water chemistries and aquarium situations, most of them are obtainable only as wild-caught, imported fish. Thus, knowledge of where and under which conditions the fish live is important in designing a strategy for their successful maintenance.

Some natural parameters

Not all South American rivers are chemically equivalent. Three major types are recognized: whitewater, clearwater and blackwater rivers.

The Amazon River is a typical whitewater river loaded with suspended sediment, usually clay, from the Andes. Because of this turbidity, the water appears white, or more accurately tan ('café-au-lait'). Because of their limited transparency, whitewaters do not support rooted plants. However, because they are nutrient-rich, extensive 'floating meadows' of such flora as water hyacinth and floating grasses arise in the calmer backwaters of the river and support large communities of insects and fish. Water parameters from a typical whitewater habitat reported by the German aquarists and fish collectors Linke and Staeck (1985), are as follows:

pH : 7.1
General hardness (GH) : 2.9 °dH
Carbonate hardness (KH) : 3.9 °dH
Conductivity : 154 microSiemens

Clearwater rivers typically drain the ancient rock massifs of Brazil and the Guianas. There are few leachable materials so the waters are very transparent. The Rios Tocantins and Xingu which arise from the Brazilian plateau near the mouth of the Amazon are two classic examples of clearwater rivers. Again, Linke and Staeck's (1985) measurements are instructive:

pH : 7.2
General hardness (GH) : 4.9 °dH
Carbonate hardness (KH) : 4.7 °dH
Conductivity : 142 microSiemens

Blackwater rivers, often tea-coloured but highly transparent, derive their coloration from humic and tannic acids leached from the flooded vegetation that grows on the white sands of the lowland flood plains through which they typically flow. Blackwater rivers, like the Rio Negro in Brazil, are usually highly acidic (pH 4-6) and are nearly as soft as distilled water. They are also the least productive. Few aquatic insects survive the extreme acidity and blackwater fish depend largely on foods that fall into the water from the forest. Linke and Staeck's (1985) measurements from a Peruvian blackwater habitat are:

pH : 6.0
General hardness (GH) : 0.12°dH
Carbonate hardness (KH) : 0°dH
Conductivity : 17 microSiemens

Cichlids also inhabit coastal rivers and lagoons, particularly those on the eastern side of South America which drain towards the Atlantic Ocean. Often these waters are moderately hard, sometimes slightly salty, and fish taken from these regions (*Gymnogeophagus* species and others) prosper in water made harder by filtration through dolomite (calcium carbonate) or by the addition of Rift Lake salt mix.

All of these water types present the stillwater (lentic) habitats preferred by cichlids. Most of these habitats have mud bottoms with abundant leaf litter and submerged trees. Sometimes the mud bottom supports anchored plants, particularly at the shoreline of lakes

Below: *A typical cross-section of a creek showing the species community, and distribution. Fishes include cichlids, characoids, rivulins and the predator* Hoplias malabaricus. *After Dr F. Froehlich.*

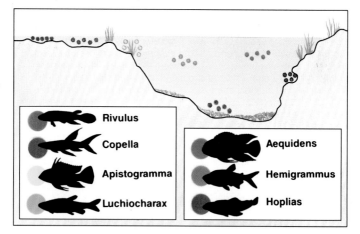

Rivulus

Copella

Apistogramma

Luchiocharax

Aequidens

Hemigrammus

Hoplias

or in swamps. There are few rocks, no gravel, and only occasional sand. While the water temperature in the main river fluctuates no more than one degree around 29°C (84°F), smaller, isolated lentic waters may vary from a low of 23°C (74°F) in small, shaded forest streams to as high as 34°C (93°F) in stagnant pools. The dissolved oxygen content, so critical to fish life, depends both on the temperature of the water and on its rate of flow, the latter often providing mixing of the surface. In the subtropical regions of Uruguay, Paraguay and Argentina, water temperatures are considerably cooler dropping down as low as 5°C (40°F) during winter.

General suggestions

Since the majority of the South American cichlids one is likely to encounter are Amazonian in provenance, this is a reasonable place to start in defining a general maintenance strategy.

Water chemistry

With the exception of species from the coastal rivers, most South American cichlids seem to prefer water whose pH is about neutral or below, and very to moderately soft. Blackwater species have more extreme requirements as already outlined. Many aquarists have tap water that is 'liquid rock', ie hard and alkaline, better suited to African Rift Lake cichlids. With the technology available today it is

Above: *Some coastal species like this Rainbow Eartheater,* Gymnogeophagus rhabdotus *from southern Brazil, appreciate harder, more alkaline water then their Amazonian relatives.*

possible to prepare water with parameters matching the particular needs of the fish you choose to keep.

One solution to softening water is the use of ion exchange resins. These resins (there are usually two of them in a mixed bed) exchange sodium (Na^+) for calcium (Ca^{++}) and magnesium (Mg^{++}) ions (cation exchanger), and chloride (Cl^-) for carbonate (CO_3^-) and sulphate (SO_4^-) ions (anion exchanger). While they soften the water it is nevertheless 'salty' with Na^+ and Cl^-. Moreover, the resins must be regenerated at regular intervals, using hydrochloric acid and sodium hydroxide, corrosive and potentially dangerous chemicals.

A better solution to creating soft water is the use of a reverse osmosis (RO) system. In reverse osmosis processing, tap water is passed under pressure through a membrane, whose pore structure is small enough to allow water, but not dissolved minerals, to pass through. The 'strained' water is collected into a sump (typically a plastic litter bin or vat). It is essentially distilled and may actually require 'hardening' by the addition of salts. RO processed water allows complete control over the composition of aquarium water and is particularly useful for delicate blackwater species. Membranes must be back-flushed regularly and replaced semi-annually as needed. Although initially expensive, RO units are well worth the investment if you are serious about keeping rare South American cichlids.

Demineralised water may also be obtained commercially in small volumes as bottled distilled water. If you are keeping a small number of dwarf cichlids in a few small volume tanks, this may be a workable solution. However, remember to add some aquarium salt (non-iodised NaCl) at a rate of 1g per 45 litres (approximately 1 level teaspoon to 10 gallons). Rainwater is not recommended because of the industrial pollutants and impurities which inevitably occur in it. If none of the above suggestions are possible, then simply select fish that are compatible with your water. There are several South American cichlids that are adaptable to conditions of harder, more alkaline water.

Often the water must be acidified. This is best done naturally using either peat or peat extract. Peat leaches natural humic and tannic acids which acidify water filtered through it. Moreover, it will also chelate toxic heavy metals. Peat can be obtained in bales or in pellets. Make sure it does not contain fungicide or fertiliser. If you don't care for the fuss and mess of preparing raw peat, commercial peat extracts are available in the trade.

Water quality and filtration

South American cichlids can be finicky about water quality. This is easily handled with today's filtration technology coupled with regular water changes.

Many of the larger cichlids are heavy and messy eaters. Clearly these require mechanical filtration to remove the large quantities of particles — uneaten food, faeces — these animals produce.

Mechanical filtration can be achieved by the use of simple floss-filled internal box filters or external power filters, as mentioned. Remember, most South American cichlids come from stillwater environments and will not appreciate strong currents. Instead of one huge power filter, use several of smaller capacity. With filter floss as a mechanical pre-filter, a foam block or chemical adsorption resin

Below: *Many of the larger cichlids like this* Aequidens tetramerus *are heavy and messy eaters and require mechanical filtration to remove the suspended particles they produce.*

can be added to effect biological and/or chemical filtration. Floss must be changed regularly as materials build up and cut down on the flow of water. In the author's opinion canister filters are not suitable for larger cichlids and should only be used in combination with other filtration methods for all but medium-sized and dwarf cichlids.

Water temperature and aeration
Another important parameter to consider in maintaining tropical South American cichlids is water temperature and its effect on dissolved oxygen content. Most Neotropical cichlids should be kept in the range of 25-29°C (78-85°F), which means that the oxygen content of the water will be relatively low (there is more dissolved oxygen in cold water than in warm). Heavy aeration with multiple air stones or internal air-driven filters will aid in mixing the oxygenated surface water, thus increasing the general oxygen content of the water. A wet-dry trickle filter in combination with aeration to boost oxygen levels. This is particularly helpful when keeping rheophilic (rapids-loving) species which depend on higher dissolved oxygen contents.

Specific set-ups
There are several general set-up strategies that seem to work well for particular kinds of South American cichlids and which provide a useful starting point. There is no 'best

Above: *Rheophilic (rapids-dwelling) cichlids, like this* Retroculus lapidifer *from the Rio Tocantins, require the higher dissolved oxygen content of wet-dry trickle filtration or heavy aeration.*

way' to handle individual species, and hobbyists will need to experiment and fine-tune these recommendations as applied to their own particular situation. Consult the **Species Catalogue** for further information.

Dwarf cichlids
These small, relatively peaceful cichlids can be housed in smaller tanks. A compatible pair of most of these species will get along in a 45-litre (10-gallon) tank provided there is adequate shelter. However, as is true for most situations, bigger tanks are better for reasons of water stability. I recommend 90 or 135 litre (20 or 30-gallon) 'longs'. Filtration by a combination of sponge, undergravel and/or canister filters is recommended with additional aeration essential because of the higher maintenance temperatures of 25-30°C (78-86°F). Many dwarf cichlids are harem polygynists — one dominant male spawning with several females — with each female holding a separate territory. Thus, adequate shelter in the form of flowerpots, or other caves is recommended. Typically shy, a well-planted tank, including rooted and floating surface plants to cut down on top

illumination, is recommended. Dwarf cichlids rarely dig. They may be kept with other peaceful 'community tank' fish like tetras, hatchetfish, pencilfish and bottom scavengers like *Corydoras* sp. catfish or small loracariids (eg *Pekoltia* sp.). The use of such dither fish helps make these retiring cichlids less shy. Attention to water chemistry (soft, acid) and water quality is essential for success with these fish as is a varied diet favouring frozen and occasional live foods.

Mouthbrooding and medium-sized Acaras

With the exception of slightly larger tanks, mouthbrooding Acaras of the genus *Bujurquina* prosper under the set-up recommended for dwarf cichlids.

Eartheaters

Eartheaters, which sift the substrate for edible detritus, have some special needs as aquarium fish. They should be kept over sand or fine gravel so they can practise what they were designed to do. However, this habit makes rooted plants an impracticality. Both Java Fern (*Microsorium pteropus*) and Java Moss (*Vesicularia dubyana*) are recommended as is some form

Below: *Most suckermouth loracariid catfish, like this 'Mango Pleco', can be kept with larger South American cichlids.*

of floating plant to cut down on surface light intensity, as these can be shy fish. Shelter in the form of a tangle of bogwood which also provides a site of attachment for the plants works well and leaches the tannic and humic acids which are welcomed by these fish. Again, tanks with larger bottom areas are prefered to 'high' tanks of a similar capacity. Water should be on the soft and acid side and kept very clean and relatively warm: 25-29°C (78-83°F). Filtration via a combination of external power filters, canister filters, and/or wet-dry trickle filters is recommended as is heavy additional aeration. The fish are fed a variety of prepared, pelleted (sinking) foods as well as frozen foods (bloodworms) and occasional live foods, particularly worms. Larger dither fish in the form of silver dollars (*Mytennis, Myleus* species) or elongated hatchetfish (*Triportheus* species) are recommended as are most catfish, including some large loracariids.

Pike cichlids

Contrary to reputation, many pike cichlids are reasonable fish for the cichlid community. They can be aggressive, but usually only with members of their same species. To this end, communities containing a number of pike cichlids should restrict the population to pikes of near identical size, and should provide adequate shelter in the form of PVC piping cut to

Above: *Characins, like this Diamond Tetra (*Moenkhausia pitteri*), make excellent dither fish for dwarf and medium-sized South American cichlids.*

appropriate lengths, one for each fish. As a group, the majority of pikes are undemanding when it comes to water chemistry or even quality. However, they are voracious and messy feeders, and good mechanical filtration, in the form of external power filters, is recommended. Live feeder fish are necessary for only the most stubborn of wild-caught specimens: most pikes can be converted to floating freeze-dried krill, frozen bloodworms even prepared foods. Tankmates, of course, should be limited to fish that cannot be swallowed: other cichlids of similar temperament and even large silver dollars work well.

Other large Neotropical cichlids

Many of the larger species, typically 'Cichlasomines', pose problems in maintenance, with the two main problems being managing aggression and maintaining water quality. There are at least two approaches to managing aggression in the large cichlid community. One is to keep them relatively crowded in tanks devoid of any shelter which diffuses the aggression amongst a large number of individuals and which provides no visual markers (rocks, driftwood, etc) for establishing territory. The result is often a surprisingly peaceful group of thugs. The second approach is to maintain several pairs of different species with adequate shelter (overturned flowerpots, etc) and tank space and let the pairs establish territories. With other pairs serving as targets, pair bonds are enhanced. And if enough room is provided, most of the aggression will take the form of ritualised display and threat with little harm done to the combatants. Such an arrangement often promotes spawning of otherwise highly aggressive, unspawnable fish. Sometimes individual rogue fish must be maintained by themselves. While the actual details of water chemistry and quality vary from species to species (see **Species Catalogue)**, one general point can be made: big cichlids make big messes! As such, mechanical filtration in the form of several external power filters and regular, large water changes are essential for the successful maintanance of the larger species. Typically these fish are omnivorous and will eat a wide variety of prepared pelleted, freeze-dried and frozen foods. Earthworms are a great treat.

Breeding South American cichlids

Of all aquarium fishes, Neotropical cichlids are perhaps the most interesting to propagate in the aquarium because of their highly developed pair bonding and parental behaviours. The majority of South American cichlids form stable monogomous pair bonds through interesting and elaborate behavioural rituals and exercise diligent, even gentle, care of their eggs and youngsters. These courtship and spawning behaviours are a wonder to behold and are one of the main reasons for keeping Neotropical cichlids in the home aquarium.

Sexing cichlids and establishing compatible pairs

The first step in spawning any fish successfully is obtaining males and females. In the case of South American cichlids, this also means obtaining **compatible** males and females, since pair bonding and co-operation is so central to reproduction in these fish, at least for the monogamous species which are in the majority. Harem polygynous species, where male-female interaction is brief and males are likely to spawn with several females in turn, pose less of a problem: a mating group of predominantly females must be provided with space and shelter.

Many South American cichlids are **sexually dimorphic**—males and females are easily discriminated on the basis of size, overall shape, finnage or colouration. But many South American cichlids are **sexually isomorphic,** that is, there are no apparent differences between the sexes. Nevertheless, many 'isomorphic' species may still be sexed on the basis of subtle differences. In general, males of most species are slightly more elongated as measured by the distance along the ventrum from the point of insertion of the paired ventral fins to the anal fin. In general, ripe females are slightly heavier through the abdomen than their consorts. However, these are differences that often only experienced aquarists can detect. The urogenital vents of males and females differ (the female's housing a blunt ovipositor, the male's a pointed tube) and in theory, they can be sexed by netting them and examining the aperture, but in practice, 'venting' is difficult and

Below: *Many South American cichlids, like these marbled pike cichlids,* Crenicichla marmorata, *are sexually dimorphic. The female, on top, has a white-edged dorsal fin and develops a cherry-red distended belly when ripe.*

Above: *Some South American cichlids, like these 'Cichlasoma' atromaculatum, are sexually isomorphic and not easily sexed. The female of the pair, in the forefront, is slightly more rotund than her consort.*

often unreliable. The proof occurs only when the tubes actually descend during spawning and the 'egglayer' is identified firsthand. For sexually isomorphic species, aquarists should purchase a group of 4-8 individuals, raise them, and allow them to pair off naturally. As females often grow more slowly and attain a slightly smaller adult size, a good mix of sizes, both large and small, is recommended in that founding group. This is good general advice for obtaining breeding stock of all South American cichlids, regardless of their sexability.

Often, however, we are not fortunate enough to have our pick of several juveniles, rather only one or a few adult specimens are available, or the price prohibits purchasing more than two. For sexually dimorphic species, in general, adult males tend to be larger and to have more filamentous finnage, in addition to whatever colorational differences (sexual

dichromatism) exist. These differences are described in the **Species Catalogue** for many species, and in some cases are so pronounced that the two sexes in question could be (and often have been) described as separate species! Unfortunately, newly-imported wild specimens are often in poor shape, emaciated, with tattered fins, and no colour. Sometimes, if you are patient and watch the fish in your dealer's tank, their interactive behaviour will suggest their sex. If two cichlids engage in lateral displays or jaw-locking, or one butts the other repeatedly in the genital region, you may be watching courtship behaviour. When all else fails, take a gamble and hope the odds are with you: buy as many as your budget allows and hope the fish will sort it out in your tank.

Assuming you have both sexes, the next task is to establish a compatible pair. Sometimes this is easier said than done, particularly if you have only two fish and must arrange a 'blind date'. If you are lucky, they will accept each other right away. The chances of this happening are increased if the two fish are added to a new (ie strange) tank together, so that neither has proprietary rights.

The use of **target fish** to help cement pair bonds is a useful strategy in breeding monogamous neotropical cichlids. Target fish are usually other cichlids, either a third conspecific, or multiple individuals of another species of similar temperament. Larger non-cichlids can often be used as well. The strategy is to provide a reason for the intended pair to work together and vent their aggression towards the other fish and not towards each other. Obviously, appropriate sized tanks with the right type of shelter should be chosen if this strategy is to work: death of the target fish is not the desired outcome.

It is often useful to separate the intended consorts for a while before they are put together. This allows them to ripen with good feeding and to 'meet' each other visually and chemically before the 'date' is arranged. They may be placed in separate tanks next to each other so they can see each other, or preferably, on either side of a porous divider in the same tank. The best divider is one which allows them to see and smell each other: plastic 'egg crate' light diffusing grating, available from most home improvement stores, fits the bill here. The 1cm (½in) lattice allows the fish to see each other and permits free exchange of water between the two compartments

Above: *Target fish are useful in helping to cement pair bonds. Here two pairs of Heckel's Threadfinned Acara,* Acarichthys heckelii, *square-off over a territorial marker.*

making filtration and heating easier and allowing the fish to sense each other chemically. With good feeding, and if they are indeed of opposite sex, the intended consorts will often begin displaying to each other across the divider, At this time the divider should be removed and the two fish allowed to encounter each other. Of course, the wise aquarist will take the time to watch that encounter—you may be called on to referee! Often the 'pair' must be separated right away if one or the other is not ready, and it may take many weeks and many tries for them to 'get it right'. The behavioural correlates of successful pair bonding usually involve a series of lateral displays where both fish approach each other, often with gills puffed out, and orient head to tail, side by side, often 'beating' the water between them. Genital butting and jaw-locking are also components of courtship behaviour. And if one of them 'blinks'—pulls back in fear or unreadiness—the tryst is over: courtship turns abruptly to aggression and the vanquished risks being liquidated if not removed.

Breeding from incompatible fish

Sometimes, they never get it right, even if they are male and female. This is often the case with large, wild cichlids. All is not lost, however. One can spawn incompatible fish using the 'incomplete divider' method. This works particularly well if egg-crate divider material is available. A suitable substrate, usually a flat stone, is placed on the bottom, under and straddling the divider. Ripe females will often lay eggs on this stone and the male can fertilise them from his side of the divider, particularly if the filter outputs are set up to provide a current that favours broadcast of the milt or sperm in the female's direction. Usually, considerably less than one hundred per cent fertilisation will be achieved, but for many of the larger cichlids whose spawns can approach 1000 in number, this is actually a blessing! Upon hatching, both parents will care for the fry, and it is wonderful to watch them swimming back and forth from parent to parent across the divider.

Another twist on this method takes advantage of the fact that many female Neotropical cichlids are considerably smaller than their male consorts. By cutting a series of holes in the divider, large enough to allow the female to pass, but too small to permit the male to follow, the female can determine her own access to the male. She simply swims to his side of the tank when she is ripe and ready. This has been called the 'hidey-hole' method.

Conditioning breeders

Assuming a compatible pair, or a mating group of appropriate sex ratio, the fish must be conditioned by the feeding of a variety of rich, often frozen and live foods. The fish should be fed regularly and often,

Right: *Some Eartheaters and Acaras, like this female Parana Port Cichlid,* Cichlasoma paranense, *choose platforms, like this water-logged leaf, to hold their spawns.*

and appropriate water quality should be maintained. It may take several weeks, or longer, to bring wild fish into ripeness.

Creating appropriate conditions

Much of the strategy for tank selection and set-up discussed previously applies here. In addition, both shelter and appropriate spawn receptacles must be provided. For cave spawners, like many of the dwarf cichlids and the pike cichlids, inverted ceramic flower pots with a notched rim or with the drainage hole enlarged should be provided. For polygynous species, one per female is required. Smooth rocks or slate are often the choice of substrate-spawning cichlids. Even waterlogged leaves, dried oak or rubber tree may be useful for Acaras (eg the *Bujurquina* sp.) that practise 'moveable platform spawning'. It is often useful to introduce the female first so that she may become familiar with the environment and stake out territory before dealing with the larger, ardent male. If target fish are used, make sure they have hiding places: PVC piping of appropriate diameter and cut to appropriate lengths provides useful escape.

The actual spawning may often be triggered by a combination of water and temperature changes. Typically, elevated temperatures will do the trick. Slowly raising the water temperature 2-3°C (4-6°F) over a period of hours or days, even as high as 32°C (90°F) for some species (be careful—watch the fish

for signs of distress and always have lots of aeration) will often trigger spawning. Combine this with partial water changes (20-50 per cent) using unaged tapwater of the same temperature. For some species, the addition of 'pure', RO processed water in modest amounts will simulate the rainy season. Again, easy does it—go slowly in changing all of these parameters and be patient! It often takes weeks of cyclical changes to stimulate a spawning. Also, some species may never spawn in captivity.

Rearing the fry

Assuming you have a clutch of eggs, the cichlid aquarist has several options. One of the joys of keeping cichlids in the first place is, of course, observing parental care. In fact, fry left with parents usually grow faster and sturdier than those removed from their care. However, parents often prove to be egg-eaters and if the species is particularly rare, intervention may be appropriate. Often egg-bearing vanishes after a few spawns and is associated with young, inexperienced or transiently infertile pairs. Sometimes, traffic by their tank will cause a shy pair to terminate care: try placing dark paper over the front glass with a small peephole for you to watch through. If egg- or fry-eating continues beyond the first few spawnings, you may elect to play foster parent. The egg-bearing substrate, assuming it is removable, may be pulled from the tank and placed in a smaller tank containing clean water siphoned from the main aquarium. Alternatively, the parents may be removed. Gentle aeration will ensure a steady supply of oxygenated water for the eggs and will keep debris from settling on them. Of course, appropriate water temperature should be maintained. Some aquarists advocate the addition of chemical bacteriacides or fungicides like Methylene Blue or Acriflavine to the water. Others prefer transfer of the eggs to fresh tapwater, insisting that the chlorine

Above: *This female dwarf cichlid,* Apistogramma agassizii, *in broodcare coloration, vigorously guards her newly-laid eggs.*

(or chloramine) acts as a disinfectant. In fact, you cannot hatch an infertile egg; however, these chemicals can keep the spread of harmful microorganisms from fungusing 'good' eggs to a minimum. These dyes are available commercially and aquarists are advised to consult other hobby books on their usage. Typically half the usual medicinal dose is recommended.

Another alternative is to remove the fry to another tank. After about four days, the eggs hatch and the larvae, called 'wrigglers' because they can not yet swim but still flex their tails rapidly in a wriggling motion, are maintained in a ball by the parents. Some wrigglers have sticky secretions produced by special glands on their heads that allow them to be attached to plants by their parents who spit them there. Parents will often excavate a series of pits and move the wrigglers, carrying them in their mouths, from pit to pit. They eventually exhaust all of their yolk, develop fins and a workable mouth and become free-swimming and capable of feeding about four days later. The actual schedule is dependent both on the species in question and the temperature of the water — the warmer the faster.

Wrigglers are quite delicate, but free-swimming fry may be carefully siphoned out with a large-diameter siphon tube. It is often a good idea to leave a few fry with the parents. Often, when a spawn is snatched, the parents will quarrel, sometimes fatally. The fry must then be fed several times a day. Fry too small to take newly-hatched brine shrimp (Artemia nauplii) may be started off on commercial liquid fry food, microworms, or encapsulated rotifers (available commercially) until such times as they can handle Artemia nauplii. These are simply hatched as per instructions in most hobby texts or on the container of eggs, which are available at most shops. With certain fish, one can make do with frozen newly-hatched brine shrimp, or even crumbled dry food, but growth rates and health are much more satisfactory with live first foods. The fry tanks should, of course, be filtered. Sponge filters from established tanks are preferred. Regular partial water changes, small but frequent, are recommended, as are the addition of snails or regular siphoning of the bottom to dispose of the uneaten and decaying shrimp.

Fry left with parents will be tended for several weeks. Typically both male and female will provide perimeter defence and herd the fry around the tank as they forage. It is interesting to watch the parents 'call' their youngsters with a flick of the ventral fins when danger approaches: they all drop to the bottom beneath the parents where

Below: Many Eartheaters, like this Red Hump female, 'Geophagus' steindachneri, are mouthbrooders and continue to offer their fry buccal protection.

they may be collected and moved if need be. It is particularly satisfying to watch parental cichlids guard and tend their fry. Fry left with parents must be fed, as explained above, when free-swimming. It is a good idea to remove other tank occupants at this time, or risk loss of the spawn. In particular, catfish will dispatch a clutch in no time after the lights go down. One solution is to leave the lights on 24 hours a day until the fry are large enough to escape predation.

In the case of mouthbrooding cichlids, like the Bujurquina species or many of the Eartheaters, fry-tending includes uptake of the young into the buccal (throat) cavity. Some cichlids, like Uaru and Discus, actually provide initial nutriment in the form of mucoid secretions on their flanks which the fry feed on (contact feeding). Of course, custodial care has its temporal limitations, typically six to eight weeks for most species, although some pike cichlids have been reported to defend 'fry' for as long as one year. Tired parents make their condition known and actually appear to avoid their youngsters. Failure to remove the fry at this point usually results in them being eaten by the parents, often as a prelude to the next spawning. Once on their own, juveniles respond positively to the conditions outlined for adults. At this point, they can be shared with aquarist friends or sold to shops or wholesalers to defray the cost of their upkeep.

Breeding aquarium fishes, and cichlids in particular, is a wonderful extension to the tropical fish hobby, accessible to all aquarists with the interest and energy to provide the appropriate conditions. You do not need many tanks and a lot of space: even the smallest of the dwarf cichlids provide the full range of cichlid parental behaviour. When cared for properly, you cannot stop fish from spawning, and that is the ultimate sign that the captive environment you have created is successful!

Species Catalogue

Nomenclature and species groupings

The taxonomy of South American cichlids is currently in a state of flux. The Swedish ichthyologist, Sven O. Kullander, began a re-evaluation of the names and placement of cichlids of the Americas in the early 1980s. His studies, to date, have concentrated on species from the Peruvian Amazon and the Guianas, but have produced results whose nomenclatural ripples have been felt throughout the previously accepted systematics of the Neotropical *Cichlidae*. His work is currently unfinished and has not yet been embraced by all the ichthyological community. Nevertheless, many of the

conclusions he has reached seem appropriate and useful and are therefore used here.

Why are scientific names important? Why should hobbyists care? Can't we get along with common names? There are several answers to these questions. Scientific names allow scientists and others to talk to each other about particular fish with precision. The latinised scientific binomial name is unique to each species and serves as a symbolic shorthand for that fish. Common names could do the same, but a standardised list of unique common names for each species does not yet exist. Many common names are used for

several different fish. For example, there are at least two cichlids with the common name *'Flag Cichlid'* in the aquarium hobby, *Laetacara curviceps* and *Mesonauta festivus*, but neither looks even remotely like the other nor are they closely related. So the latinised scientific names are the only way to sort out what you actually have.

A second reason why scientific names are useful is that they attempt to express something about evolutionary relationships. For example, most aquarists would be easily persuaded that the many elongate, torpedo-like pike cichlids are closely related and may have arisen from a common ancestral piscivorous cichlid which had this basic body plan. Placement of the majority of pike cichlids in the genus *Crenicichla* attests to this relationship. Some of the pike cichlids have a reduced snout and consistent changes in the arrangement of their teeth. These have been placed in the genus *Batrachops*. Such placement suggests that these species are more like each other and form a natural sub-grouping within the greater set. Ichthyologists may debate that placement or suggest that the changes are too unimportant to warrant splitting the subgroup from the main assemblage (in fact, Kullander has

lumped *Batrachops* together with *Crenicichla*!) Nevertheless, it is both interesting and important to realise that these cichlids have solved similar ecological challenges in similar ways, and are presumably derived from a common ancestor. One can make too much of the evolutionary (phylogenetic) implications of pigeonholing like species in the same genus. However, often natural groupings do speak volumes about the evolutionary processes of radiation and speciation. In a more practical sense, species-groupings suggest commonalities in aquarium care and spawning to aquarium hobbyists attempting to maintain and captively-breed fishes. Many of the 'species-groups' make intuitive sense to aquarists who have actually worked with these fish in the aquarium, rather than simply studying pickled examples.

Above: Dicrossus maculatus *is a species which has been newly re-imported from the Brazilian Amazon.*

For these reasons, I believe that Kullander is on the right track with his re-evaluation of the Neotropical *Cichlidae.* Thus, the scientific nomenclature used throughout this species catalogue reflects the current state of this re-evaluation. I have tried to explain the reasons for the rearrangements to avoid undue confusion, and have tried to indicate the 'pre-Kullander' names under which information about these fish can be accessed through earlier aquarium literature. I have also presented the species in what I believe are their natural groupings as aquarium fish which, more often than not, reflect their actual evolutionary links. These major groups include the Acaras, the

Eartheaters, the Apistos and allied dwarf cichlids, the pike cichlids, the Cichlasomines and derivatives, and finally, a few miscellaneous species which seem to fit nowhere and which are probably transitional in an evolutionary sense. In certain cases, assignment of an aquarium fish to a definite species is difficult and tentative: many may well be new to science and undescribed. These fish are referred to as 'sp. affin.', short for *species affinis* which in scientific circles means "looks like, has affinity with, a species of this name but we cannot be sure". There are several entries in this catalogue bearing that designation. My coverage in this catalogue is nowhere near exhaustive, which is impossible for a book of this scope. However, species which can be expected to be encountered in the aquarium trade have been chosen as representatives of their particular species group.

The Acaras

This group of South American cichlids is the most primitive in terms of generalised body structure and ecology. Acaras are typically egg-shaped omnivores which, with certain exceptions, are monogamous, biparental substrate spawners. In the aquarium, they are amongst the least demanding of the South American *Cichlidae* with regards to water chemistry and quality or dietary requirements. They are also the least likely to enter the hobby as anything other than contaminants and are usually sold under the catch-all name 'Port Cichlid' or simply 'Acara'. In fact, this assemblage is rather specious and heterogeneous.

Heckel, in 1840, created the genus *Acara* for these fish, choosing the native Guarani name for these cichlids. Eigenmann & Bray, in 1894, put forward the genus *Aequidens* as a replacement for *Acara* when they discovered that the type specimen, *Acara crassipinnis*, was actually a junior synonym of the oscar, *Astronotus ocellatus*, a fish substantially

different from most other Acaras. The name *Aequidens*, which translates as 'equal tooth', refers to the absence of enlarged pseudocanine teeth which are often found in members of the Cichlasomine lineage. All members of the genus *Aequidens* have, instead, small conical teeth and most, but not all, have only three hard rays in their anal fins compared with four or more rays in the 'Cichlasomines'. Most of the Acaras have been written about in the hobby literature under the name *Aequidens*.

In fact, the genus *Aequidens* is a mixed collection of fishes. Kullander has redefined the genus *Aequidens* and has restricted it to the larger forms of Acaras. These we will call the 'true' Acaras. He has also erected the genera *Bujurquina*, *Laetacara*, *Krobia*, *Cleithracara*, and *Guianacara* to hold most of the other 'orphaned' Acaras. Those that he has not yet dealt with and which have not yet been formally reassigned are designated for now simply *'Aequidens'*, in quotation marks.

TRUE ACARAS

The species roster of True Acaras (*Aequidens*) includes *chimantanus*, *diadema*, *metae*, *pallidus*, *paloemeuensis*, *patricki*, *plagiozonatus*, *potaroensis*, *tetramerus*, *tubicen*, *uniocellatus*, *viridis*, and several as yet unnamed species. Note that the aquarium fish *Aeq. awani* is in fact, *Aeq. viridis*. Some of these species, particularly *Aeq. tetramerus*, are quite cosmopolitan in their distribution and exist in several colorational/geographical varieties. Only a few of these species have been imported, most as accidental 'contaminants' of other cichlid shipments. Most of these species grow quite large, about 25-30cm (10-12 in), and are somewhat belligerent. All but *Aeq. diadema*, a primitive mouthbrooder, are biparental substrate spawners. Despite their belligerence, most are easily cared for in the aquarium.

Aequidens metae
Rio Meta Acara
- **Distribution:** Rio Meta, Colombia
- **Length:** 20-30cm (8-12 in) in captivity
- **Diet:** Undemanding omnivore
- **Sexing:** Essentially isomorphic. Males slightly more elongate than females
- **Aquarium maintenance and breeding:** Although beautiful, *Aeq. metae* can be quite belligerent. Best suited for the mixed, 'rough' cichlid community. Undemanding with respect to water chemistry and quality, as well as feeding. Although growing quite large, these are precocial spawners which will breed at 10cm (4 in) or less. At this

Above: Aequidens metae *is an attractive 'True Acara' from Colombia.*

size they handle easier. Typical biparental substrate spawners.
- **Comments:** *Aeq. metae* is relatively uncommon in the hobby. It was originally described from the Rio Meta, a tributary of the Rio Orinoco, and is found as an occasional juvenile accidental contaminant in cichlid shipments from Colombia. More recently, it has been available as tank-raised juveniles from Europe.

Below: Aequidens diadema *is the only known mouthbrooding 'True Acara'.*

BLUE ACARAS

The Blue Acaras have yet to be generically reassigned, hence the use of quotation marks around the old genus name. This group of medium- to large-sized Acaras contains some of the more beautiful members of the Acara lineage. Species include *'Aeq.' biseriatus, coeruleopunctatus, latifrons, pulcher, sapayensis, rivulatus* and the *'rivulatus'* complex. Several of these, *Aeq. pulcher* and *Aeq. sp. affin. rivulatus*, have become staples of the aquarium trade and are commercially propagated in Asia and Florida.

'Aequidens' pulcher
Blue Acara

● **Distribution:** Northwestern South America, Trinidad
● **Length:** 15-20cm (6-8 in) in the aquarium
● **Diet:** Undemanding omnivore
● **Sexing:** Essentially isomorphic. Males slightly more elongate than females
● **Aquarium maintenance and breeding:** Relatively undemanding with respect to water chemistry, quality and overall maintenance. They can be somewhat belligerent, so adequate shelter and appropriate tankmates should be provided. They are ready biparental substrate spawners, and will lay 100-500 eggs on rocks or other hard substrates, and make exemplary parents, even at 5-7cm (2-3 in) size.

● **Comments:** *'Aequidens' pulcher* hails from northwestern South America, the coastal regions of Venezuela including the island of Trinidad, down to the Orinoco drainage. However, since it has been commercially bred in Asia and Florida we rarely see wild specimens of this beautiful fish. *'Aequidens' pulcher* is replaced by the closely related *'Aeq.' latifrons* in northern Colombia, a higher-bodied form with more iridescent scalation. *'Aequidens' coeruleopunctatus* replaces *'Aeq.' pulcher* at the Colombian-Panamanian border and is found as far north as southern Costa Rica in Central America (see page 77). Care is identical for all. Both *'Aeq.' pulcher* and *'Aeq.' coeruleopunctatus* have proved to be movable platform spawners in the wild habitat, using waterlogged leaves as the preferred egg receptacle.

Below: 'Aequidens' pulcher, *the Blue Acara, is an undemanding mid-sized Acara.*

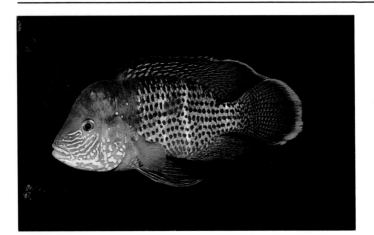

'Aequidens' sp. affin. rivulatus

Green Terror, Rivulatus

● **Distribution:** Eastern Ecuador.
● **Length:** Males can exceed 30cm (12 in) in captivity, females up to about 20cm (8 in)
● **Diet:** Omnivorous
● **Sexing:** Mature males develop conspicuous nuchal humps. Females stay one-third smaller and are not as elongate
● **Aquarium maintenance and breeding:** A fairly belligerent fish that is appropriate in mixed communities of rowdy cichlids. Relatively undemanding as to water and dietary requirements. These are precocial spawners who reach sexual maturity at 10-15cm (4-6 in). At this size, they are much less liable to liquidate each other in the process of establishing a pair bond. Large specimens can be difficult to pair. Reproduction as for most biparental substrate spawning cichlids.
● **Comments:** The real *'Aequidens' rivulatus* hails from the Pacific coast of Ecuador, Venezuela and Colombia, and is rarely imported. The hobby *'rivulatus'*, the Green Terror, was first imported in the early 1970s along with *'Cichlasoma' festae*, the 'Red Terror' from eastern Ecuador. In the real *'Aeq.' rivulatus*, the centres of the flank scales are iridescent green, whereas in the Green Terror, it is the edges of the

Above: *This male Green Terror,* 'Aequidens' sp. affin. rivulatus, *lives up to its common name.*

scales, leaving the centres dark. *'Aequidens' sp. affin. rivulatus* may, in fact, be either *'Aeq.' azurifer*, or *'Aeq.' aequinoctalis*, two species earlier synonymised with *'Aeq.' rivulatus*. Two colour varieties of Green Terror are known: those with white edging to their tail and unpaired fins, and those with orange or red. They are apparently the same species. There may be several dwarf species of the *rivulatus* — complex, that grow no larger than 13cm (5 in).

SMILING ACARAS
The Smiling Acaras are a group of mostly smaller forms which share a peculiar snout marking. The genus name, *Laetacara*, derives from the Latin *laetus*, meaning 'happy'. The 'smile' consists of a series of three dark stripes extending from the eyes to the lips. The species include *L. curviceps, dorsigera, flavilabrus, thayeri*, and a number of as yet to be described forms. With the exception of *L. thayeri*, the giant of the group at 15cm (6 in), the Smiling Acaras remain small at 7.5-10cm (3-4 in) and are perfect for the planted community tank. All are biparental substrate spawners.

Laetacara curviceps
Curviceps, Flag Cichlid

- **Distribution:** Amazon drainage
- **Length:** Males up to 7.5cm (3 in), females 1cm (½ in) smaller
- **Diet:** Requires frozen or live foods for optimal conditioning
- **Sexing:** Males develop a convex head profile. In some populations, females may have one or more large ocellated spots in their dorsal fins
- **Aquarium maintenance and breeding:** These fish are best treated like dwarf cichlids of the genus *Apistogramma*. Water should be soft and acid, and kept scrupulously clean and warm at 25-29°C (78-84°F). The addition of peat or peat extract to simulate blackwater is helpful. These are excellent community tank residents which prefer planted tanks. Additionally, dither fish in the form of small schooling tetras will make them feel at home. Biparental substrate spawners whose spawns number 100-300. The fry can be somewhat small, so they should be provided with liquid fry food, microworms or rotifers before switching to newly-hatched *Artemia* nauplii.
- **Comments:** An excellent choice for the small, planted community tank. Several colour varieties are available, including blue and red

Above: *The 'smile' is easily seen on this* Laetacara thayeri, *the giant of the group.*

morphs. In addition, a second look-alike species, *L. dorsigera*, is available from the La Plata system (Rio Paraguay/Rio Parana), often in wild shipments from Argentina. Although superficially resembling *L. curviceps*, this fish develops a bright maroon to fire engine red breast and face when brood tending. Care like *L. curviceps*, however; lower temperatures are possible given the subtropical origins of this fish. A third, as yet undescribed species, a mint-green curviceps-like fish from the Mato Grosso region of Brazil is also occasionally imported, as is the slightly larger form *L. flavilabrus* from Peru, and the 'giant curviceps', *L. thayeri*, from Brazil. Care for all 'Smiling Acaras' as for *L. curviceps*.

Below: Laetacara curviceps *is the most common of the 'Smiling Acaras'.*

MOUTHBROODING ACARAS

The Mouthbrooding Acaras, *Bujurquina* species, are a group of 17 species which practise delayed (primitive) mouthbrooding. Eggs are laid on a substrate, often movable, and are fanned and guarded for 24-48 hours. Then the larvae are chewed from their eggshells and uptaken into the mouths of one or both parents where they are incubated for a further 2-6 days. The free-swimming fry commence foraging immediately and, if threatened, stream back into the open mouths of the parents. This intense parental care is practised for 6-8 weeks post-spawn.

These cichlids are elongate medium-sized at 10-15cm (4-6 in), and often have a fringed caudal fin. In addition, they all have an obliquely-oriented black lateral band that runs from the eye back to the insertion of the soft dorsal fin, the latter visible even at small size. Many of the species are found in rather restricted areas which are not commercially collected. At least three species are sporadically available: *B. mariae, syspilus,* and *vittatus.* All are excellent for the planted community tank. The name *Bujurquina* (pronounced *boo-her k'eye na*) was created from the native name for these fishes, *bujurqui.*

Above: Bujurquina mariae *is one of the more common of the Mouthbrooding Acaras.*

Bujurquina vittata
'Paraguayensis', Paraguay Mouthbrooder
● **Distribution:** Rio Parana/Rio Paraguay of the La Plata system (Paraguay, Argentina)
● **Length:** Medium-sized 10-13cm (4-5 in) Acara
● **Diet:** Frozen and live food supplements recommended
● **Sexing:** Essentially isomorphic
● **Aquarium maintenance and breeding:** Should be maintained as for the Smiling Acaras. These make good denizens of the medium-sized planted community tank as they are

Below: The 'Paraguensis', Bujurquina vittata, *is a mouthbrooding Acara from Paraguay and Argentina.*

somewhat shy. Can take lower temperatures. Spawns number 50-100 eggs which are uptaken by both parents. The free-swimming fry may need smaller first foods.

● **Comments:** Originally known as *Aequidens paraguayensis* in the hobby, Kullander relegated that name to junior synonymy with *B. vittata*. However, when sporadically available, the fish is still sold under the name 'Paraguayensis'. In the wild, these and other *Bujurquina* species preferentially select waterlogged leaves as movable platforms for their spawns. When danger threatens or if water levels drop, they simply drag the spawn-bearing platform elsewhere.

Below: *The Keyhole Cichlid,* Cleithracara maronii, *is a good candidate for the planted South American community tank.*

Above: *This* Krobia guianensis *from the Guianas is usually sold as 'Itanyi' or the Dolphin Acara.*

MISCELLANEOUS ACARAS
The remaining Acaras have been installed in a variety of genera, several with only one species (monotypic). These include *Cleithracara, Guianacara, Krobia, Nannacara* and *Tahuantinsuyoa.* Several of these are common aquarium fish.

Cleithracara maronii
Keyhole Cichlid
● **Distribution:** The Guianas (Rio Maroni in Surinam)
● **Length:** Large males up to 13cm (5 in), females a little smaller
● **Diet:** Require frozen or live food for optimal conditioning

- **Sexing:** Essentially isomorphic
- **Aquarium maintenance and breeding:** Best handled like *Laetacara curviceps* (above). Shy almost to a fault, these fish require planted tanks with plenty of shelter to feel comfortable. Dither fish, in the form of peaceful schooling tetras, recommended. A rich diet of frozen and live foods will bring them to ripeness. Moderately difficult to spawn: notorious egg eaters. Artificial incubation recommended for pairs which continually eat their spawns.
- **Comments:** The genus *Cleithracara*, from *kleithron* (Greek for lock), commemorates the common name of this fish. Commercially bred in Asia, wild specimens are rarely imported.

Nannacara anomala
Golden-Eye Dwarf Cichlid
- **Distribution:** The Guianas, principally imported from Guyana
- **Length:** Males up to 7.5cm (3 in), females half that length
- **Diet:** Undemanding omnivore
- **Sexing:** Highly dimorphic. Males have high, rounded foreheads and iridescently spangled flanks. Females remain brown and express a parallel two-stripe flank marking
- **Aquarium maintenance and breeding:** Truly a beginner's cichlid. Undemanding in terms of maintenance and diet: can be conditioned on prepared foods if necessary. Small size makes them a natural choice for planted community tanks. Several hundred eggs are laid on a receptacle and guarded assiduously by the female. When brood-tending, the parallel two-stripe pattern of the female is replaced by a distinctive latticework pattern: vertical bars intersecting the parallel stripes herald motherhood! The female is exceedingly intolerant of the male and will chase him off. In small tanks, the male may be killed and should be removed. The female takes exceptional care of her offspring who will grow even on crushed dry food. A great choice for anyone wishing to gain experience in breeding substrate-spawning cichlids.
- **Comments:** *Nannacara anomala* is usually available at low cost as the principal species in boxes of 'Assorted Dwarf Cichlids' from Guyana. There are several colour geographical variants available, including orange, black and red. A second species, *N. aureocephalis*, with a gold head and bigger body, has been described as from French Guiana, but it may well prove to be *N. anomala*.

Below: *The Golden-Eye Dwarf, Nannacara anomala, (male pictured) is a wonderful beginner's cichlid.*

Left: Guianacara sp. affin. geayi *is an interesting cave-spawning cichlid.*

second, undescribed 'geayoid' from Venezuela, recognisable from the bright orange-red spots on its cheeks (preoperculum), is also available in the hobby.

Guianacara sp. affin. geayi
'Geayi', Bandit Cichlid
- **Distribution:** Guianas, Venezuela
- **Length:** Males 15-18cm (6-7 in), females one-half to two-thirds as large
- **Sexing:** Males develop a large, block-like nuchal crest
- **Aquarium maintenance and breeding:** The requirements of *G. sp. affin. geayi* in the aquarium are few and they do well in the mixed, rowdy cichlid community tank. They are found in rocky pools in nature and are cave spawners in the aquarium. Thus they should be provided with some sort of cave, particularly a ceramic flowerpot with a drainage hole carefully enlarged to allow entrance. Eggs, often green, are laid on the vertical sides of the pot. Care of fry as usual. They will spawn precocially at a length of 75cm (3 in).

- **Comments:** 'Geayi' (pronounced *gay'-eye*) has had a rather uncertain taxonomic history owing to its transitional status. Originally installed in the genus *Aequidens*, *geayi* was moved first to *Acarichthys* (with *heckelii*) and most recently to its own newly-created genus *Guianacara*. Kullander, who did the moving, has described a series of species, in addition to *geayi*, which he has installed in the genus and which are distinguishable primarily on the basis of banding. Apparently the true *G. geayi* is found only in French Guiana, thus is unlikely to be the hobby fish whose origin was Guyana. The common 'Geayi' is probably *G. sphenozona*, but the name *'geayi'* persists in the trade. A

EARTHEATERS AND ALLIES
Few natural groupings of species provide the range of aquaristic challenge that the South American Geophagines or Eartheaters do. The genus *Geophagus* was 'created' by Heckel in 1840 for cichlids which shared a lobed first gill arch and small 'rakers' on the margin of their gills. These gill rakers allow for their habit of sifting the substrate for detritus. Mouthfuls of sand or mud taken by the elongate, underslung mouth are strained through the rakers, the food retained, and the inedibles passed out via the gill covers. Even the natives refer to them as 'Eartheaters'. The original genus *Geophagus* has recently been limited by Kullander, and the remaining Geophagine cichlids placed in other genera including *Gymnogeophagus*, *Satanoperca* and *Biotodoma*. Several species await reclassification into their own new genera, and these are designated as *'Geophagus'* species which are pending reassignment to new genera.

Within the assemblage there are three major modes of reproduction: simple substrate spawning; primitive (delayed) mouthbrooding, as in the Mouthbrooding Acaras; and immediate (advanced) mouthbrooding. In immediate mouthbrooding, pairing is brief and the female gathers the eggs, which are fertilised in her mouth, and incubates them to full term, much like her African counterparts from the Great Rift Lakes. Eartheater cichlids come in all shapes and sizes, in all colours, all dispositions, and all degrees of challenge.

Satanoperca leucosticta
Demonfish, 'Jurupari'
- **Distribution:** Guyana, Amazonia
- **Length:** Up to 30cm (12 in)
- **Diet:** In nature, these fish sift benthic invertebrates and plant material ('grut') from the substrate. In the aquarium they are omnivorous, but require frozen and live foods for optimal conditioning. They prefer sifting over fine gravel or sand. Sinking or pelleted foods recommended
- **Sexing:** Essentially isomorphic. In the wild, the male of the pair can usually be identified as the larger fish (Cichocki 1976, Lowe-McConnell 1969)
- **Aquarium maintenance and breeding:** Like most *Satanoperca* species, *S. leucosticta* demands clean, warm water, around 25-29°C (78-85°F). Demonfish prefer soft, acid water and benefit from the addition of peat filtration or peat extracts. A tangle of bogwood and non-anchored or floating plants together with dither fish will make them feel less shy. While other cichlids may be housed with them, these should be peaceful: *S. leucosticta* may be inhibited from spawning in the presence of other cichlids.

Satanoperca leucosticta is a biparental, delayed mouthbrooder. It is moderately difficult to induce to spawn in the aquarium. In the wild, pairs have been observed choosing movable platforms such as waterlogged wood or even sneakers, which they tug around with them in response to potential predation. In the aquarium, they will accept stones. About 150-300 eggs are laid and then covered with a thin layer of gravel or sand. The eggs are fanned and guarded for about 48 hours at which point the parents chew the larvae out of their eggshells and uptake them for further incubation in their mouths. Both parents participate in this buccal incubation, taking turns and passing the larvae back and forth for several days. When free-swimming, the fry are released to swarm and forage, but the parents continue to provide buccal shelter for several weeks thereafter. When alarmed, one or both parents will assume a head-down, mouth-extended posture and the fry respond by swarming to and diving into the open mouth for protection. When danger is past, the fry will be spat out or blown backwards through the gill covers (opercula) and resume their foraging. After about 3-4 weeks, the fry and parents should be separated. The free-swimming fry eagerly eat newly-hatched *Artemia* nauplii as their first meal, but grow somewhat slowly.
- **Comments:** Known in the hobby as *Geophagus jurupari*, the silvery,

Below: Satanoperca leucosticta, *the 'Jurupari' of the hobby.*

spot-faced 'Jurupari' was more correctly identified as *Satanoperca leucosticta* by Kullander in 1986. The true *S. jurupari* is a plain silver or gold eartheater with an unspotted face and is not commonly available in the trade. There is at least one other unspotted 'juruparoid' eartheater — *S. pappaterra* — from the Brazilian Mato Grosso that can be distinguished from these other two by virtue of its gold base colouration and distinctive longitudinal black/brown stripe that extends from the eye to the tail. All are maintained as above.

The name 'Jurupari', applied both commonly and scientifically to these fishes, is a native Tupi name for a feared forest demon. The natives called these fish 'juruparipindi', or 'demon's lure' (fishhook), but the relationship of this fish to the Jurupari myth is obscure. Kullander (1986) commemorated this relationship by resurrecting the genus *Satanoperca* Guenther, 1862, which translates from the Latin as 'Satan's Perch'. Hence the common name 'demonfish' for this group of interesting cichlids.

Satanoperca daemon
Daemon, Three-Spot Demonfish
● **Distribution:** Colombia: Venezuela; Rio Negro, Brazil
● **Length:** Up to 30cm (12 in)
● **Diet:** Sifting detritivore, handled like *S. leucosticta*
● **Sexing:** Essentially isomorphic. Both sexes with multiply-filamentous dorsal fins

Above: *The Three-Spot Demonfish,* Satanoperca daemon, *sports a diagnostic caudal ocellus.*

● **Aquarium maintenance and breeding:** Care is identical to that of *S. leucosticta* with a little more attention to water chemistry. *S. daemon* is notoriously difficult to rear to healthy adulthood, and has been successfully propagated in captivity only a few times. These fish are particularly susceptible to 'Neotropical Bloat', a terminal condition that involves extreme blockage and bloating of the belly region. The origin of the condition is unclear; however, strict attention to water quality is essential. Dissolved oxygen may be an important factor and wet/dry trickle filters are recommended. Spawning has been achieved in simulated blackwater having a pH of 4.5 and no measurable hardness; the use of RO — processed water is recommended. Although all *Satanoperca* species are believed to be biparental, delayed mouthbrooders, *S. daemon* may not be. In one of the very few spawning accounts published, Eckinger (1987) reports modified substrate spawning.

● **Comments:** *S. daemon* is easily recognisable by the white-ringed black ocellus on the caudal peduncle, and the two mid-lateral blotches, one in the centre of the fish just below the lateral line, the second equidistant between it and the ocellus. However, there are two other 'Spotted-Juruparoids' that offer some confusion. One of these is the newly-described *S. lilith* that enters the hobby as a rare contaminant from the Brazilian Amazon. This fish resembles *S. daemon* in all ways except that it has a single mid-lateral blotch located on and just above the lateral line. A more-commonly encountered, but nevertheless rare, lookalike from Brazil is *S. acuticeps*. *S. acuticeps* is distinctive for the three equally-spaced black mid-lateral blotches that grace its flanks and the presence of a simple black blotch, rather than an ocellus, on the caudal peduncle. All three species develop spectacular long red, multiple (five) filaments on their dorsal fins.

Neither *S. lilith* nor *S. acuticeps* have been spawned in captivity, nor are they regularly available; however, they are sometimes mixed in with *S. leucosticta* in wild shipments from Brazil. Care and approach for all should be like that of *S. daemon*.

Geophagus proximus
'Flag Tail' Surinamensis
- **Distribution:** Amazon Basin
- **Length:** Up to 30cm (12 in)
- **Diet:** Omnivorous
- **Sexing:** Essentially isomorphic

Below: Geophagus proximus *is one of the 'Surinamensoid' Eartheaters.*

- **Aquarium maintenance and breeding:** In general, the parameters established for *S. leucosticta* apply equally well here. *Geophagus proximus*, and other related 'surinamensoids', are somewhat more aggressive cichlids which can be housed and bred in a community situation. They are best kept as a group of several individuals to diffuse the aggression. Although water quality must be maintained, water chemistry seems less important than for the Satanopercoids.

G. proximus is an advanced mouthbrooder with the female uptaking the eggs. The male may or may not participate in the rearing and defence of the fry. Newly-hatched *Artemia* nauplii are a fine first food, and growth is rapid. They are reproductively competent at a size of 11.5cm (4½ in), at an age of 1-1½ years.

- **Comments:** This fish, and several related species, have been known in the trade for years as *Geophagus surinamensis* — a fish initially believed to be cosmopolitanly-distributed with a number of geographic colourational morphs. However, aquarium-based observations of spawning behaviour have suggested that several discrete species make up the 'surinamensoid complex'. While the black-chinned Guyanese form,

Above: *This male Red Hump Eartheater*, 'Geophagus' steindachneri, *has a conspicuous red nuchal hump.*

subsequently named
G. brachybranchus by Kullander, is a delayed mouthbrooder, several other Amazonian forms (*G. altifrons, megasema*), including *G. proximus*, have proven to be immediate mouthbrooders and one, the recently-described *G. argyrostictus* from the Rio Tocantins, is even a simple non-mouthbrooding substrate spawner. Although readily distinguishable on the basis of colour pattern, *G. proximus* and most of the other surinamensoids, are still sold as *G. surinamensis* in the trade. The true *G. surinamensis*, from Surinam, has yet to be imported commercially.

'Geophagus' steindachneri
Red Hump Eartheater
● **Distribution:** Colombia, Venezuela
● **Length:** Males up to 15cm (6 in), females half to two-thirds the size
● **Diet:** Omnivorous
● **Sexing:** This fish is named after one aspect of its well-defined sexual dimorphism: the large, red nuchal hump that mature, dominant males sport on the top of their heads. Females lack any hint of the hump (see photo on page 31). In addition, females remain half–two-thirds the size of their consorts, and are usually drably coloured. Males typically have some modest

iridescent spangling (green, black, orange) on their flanks. Sexually precocious at small size
● **Aquarium maintenance and breeding:** Although Red Humps, particularly dominant males, can be aggressive, these are excellent beginner's cichlids. They are totally undemanding in dietary and water requirements and will spawn at small size (females 2.5-4cm (1-1½ in). These are immediate, maternal mouthbrooders. Courtship is brief and pair-bonding non-existent. Ripe females select and clean off suitable substrates to hold their spawns and then solicit the male to participate. Eggs are laid a few at a time with the female backing up to scoop them up in her mouth. The male spreads his milt over the substrate and, presumably, sperm is picked up by the female when she scoops up the next batch of eggs. As many as 150 eggs may be laid, dependent on the size of the female, but spawns typically number 30-50. Ovigerous females are excellent single parents and rarely lose a spawn. They should be moved to separate quarters for release of the free-swimming fry 8-10 days post-spawning. As they do not eat during incubation, the females should be kept with their brood and fed for several days before being returned to the colony. Males are harem polygynists capable of spawning with several females in succession, which are best maintained as a large harem. Females kept as a harem

will ripen together and spawn nearly synchronously. Fry are large and easy to raise, even on crushed dry or prepared food.

● **Comments:** Regrettably, this fish is available in the hobby under an assortment of incorrect names, including *'G.' pellegrini* and *'G.' hondae*. *'G.' pellegrini* is the valid name of a larger, humped eartheater that hails from the Pacific slope of southwestern Colombia up to Panama, and which has rarely been imported into the hobby. *'G.' hondae* is a junior synonym of *'G.' steindachneri*, and is therefore incorrectly used. A third humped species, *'G.' crassilabrus*, is found in Panama, and has entered the hobby only through the personal efforts of dedicated amateur aquarists who have collected it there. Like *'G.' pellegrini*, it is a larger species reaching 25cm

Below: *The Mother-of-Pearl Eartheater 'Geophagus braslliensis, is an easy, beginner's cichlid.*

Above: *The Thick-Lipped Eartheater 'Geophagus' crassilabrus, is a rare 'Humped' Eartheater from Panama.*

(10 in) in length. All three 'Humped' Eartheaters are sexually dimorphic harem polygynists, and all three are immediate mouthbrooders. Since *'G.' steindachneri* occurs in several colour varieties (black, orange, green) and since neither *'G.' pellegrini* nor *'G.' crassilabrus* are not regularly imported, the 'Red Hump' you find in the trade is most likely to be *'G.' steindachneri*.

'Geophagus' brasiliensis
Mother-of-Pearl Eartheater
● **Distribution:** Southern Brazil, Argentina
● **Length:** Males up to 25cm (10 in), females half to two-thirds size
● **Diet:** Omnivorous
● **Sexing:** Mature males develop pronounced nuchal humps, females do not. Sexually precocious and sexable at small size

● **Aquarium maintenance and breeding:** *'Geophagus' brasiliensis* is a beautiful cichlid with modest requirements. Hailing from subtropical South America, these fish can withstand temperatures of 10°C (50°F) while tolerating more tropical aquarium conditions. Their coastal distribution suggests that harder, alkaline water is to their liking and the addition of crushed dolomite or oyster shell to their filter box will aid in increasing the carbonate hardness and buffering the pH. They can be conditioned on a diet of good prepared foods, although this is not advised. They are typical substrate spawners laying several hundred eggs and are exemplary parents. Fry are easily raised on a diet of crushed dry food and newly-hatched *Artemia*. Growth is rapid. A good beginner's first experience in breeding a typical substrate spawning cichlid.

● **Comments:** The Mother-of-Pearl cichlid is a truly beautiful fish. Each scale centre is marked with a blue, green or silver nacreous spot on a base colour that ranges from brown to mahogany or bright red, while the unpaired fins are spotted and striped with hyaline dots. Mature males, which can easily reach

25cm (10 in) in captivity, are truly breathtaking. Unfortunately, juvenile *'G.' brasiliensis* are brown and fail to develop adult colouration until they grow to nearly 7.5-10cm (3-4 in). This, coupled with their ease of propagation and the size of the spawns, has spelled commercial doom for this fish. Although inexpensive, they are well worth the space and attention in even the advanced cichlid hobbyist's collection for their sheer beauty.

'G.' brasiliensis is highly polymorphic. Many of the differently coloured populations may well prove to be valid species in their own right. One fish that has been confused, historically, with *'G.' brasiliensis* is *Gymnogeophagus gymnogenys*. Although earlier pictures suggested an elongated, 'brasiliensis'-like fish, more recent collections in southeastern Brazil have yielded a pearl-scaled eartheater that is a delayed mouthbrooder, the 'real' *G. gymnogenys*. This latter fish is available sporadically in the hobby and is of particular interest to cichlid aficionados for its beauty and rarity.

Gymnogeophagus balzanii
Paraguay Eartheater, 'Balzanii'
● **Distribution:** La Plata Drainage, Argentina; Paraguay
● **Length:** Males up to 20cm (8 in), females half to two-thirds the size
● **Diet:** A snail crusher in the wild. Omnivorous in the aquarium

Below: *Gymnogeophagus gymnogenys, a delayed mouthbrooder, has often been confused with 'G' brasiliensis'.*

● **Sexing:** Dramatically dimorphic. Dominant, sexually mature males, as small as 5cm (2 in), develop huge nuchal hoods on their forehead giving the fish a 'blockheaded' look. Males also express 4-5 parallel metallic blue longitudinal stripes on their flanks and have large, fan-like ventral fins spotted in blue. Females lack the spangling, have smaller yellow ventral fins, remain half to two-thirds the size of males and never develop the hood. The size differential is apparent early on

● **Aquarium maintenance and breeding:** *Gymnogeophagus balzanii* is the canary of the Neotropical cichlid world: water must be kept scrupulously clean. When water quality declines they are particularly susceptible to neuromast erosion or pitting, known as 'Head Hole'. Hailing from subtropical regions of Paraguay and Argentina, they can take temperatures down to 15-17°C (low 60s°F), but are best maintained at 24-27°C (76-80°F). Regular prepared, frozen and live foods keep these eager eaters in the best of health. They can be aggressive with each other and with other cichlids.

G. balzanii is a harem polygynist best maintained as a colony of one male and two or more females.

Above: *Male* Gymnogeophagus balzanii *is notable for the huge nuchal hump it develops when sexually mature.*

Ripe females establish and defend territories with a spawn receptacle as the focal point. About 24 hours prior to egg-laying, they assume brood care colouration which consists of a dark mid-lateral blotch, bandit eye-cheek stripe, and blackening of the edges of the ventral fins. The male's participation is brief: these are delayed, maternal mouthbrooders. The free-swimming fry are too small to take newly-hatched *Artemia* and must be offered liquid fry food, microworms or rotifers for the first few days. Growth is rather slow. Sexual maturity is reached at a size of 5-7.5cm (2-3 in) which is attained in about one year.

● **Comments:** The genus *Gymnogeophagus* was established by Ribeiro in 1918 to accommodate eartheater species lacking cheek scalation. Gosse, in 1975, revitalised the genus and described several further characters which distinguish it from the other geophagines. There are currently eight species in this genus of 'naked eartheaters', including both substrate spawners (see Gg. rhabdotus, page 21) and delayed mouthbrooders.

Biotodoma cupido
Cupid Cichlid

- **Distribution:** Amazon drainage
- **Length:** Up to nearly 15cm (6 in)
- **Diet:** Requires frozen and live foods
- **Sexing:** Reportedly dimorphic with respect to the iridescent blue, vermiform markings that develop on the face of mature specimens. In males, these are lines, whereas females have spots. Otherwise identical in size and finnage
- **Aquarium maintenance and breeding:** A somewhat delicate fish best treated like the *Satanoperca* species. Keep them warm at a temperature of 27-29°C (80-84°F), clean, and in soft, acid water (pH 5-6, less than 1°dH) — fine in a planted tank. They tend to be somewhat scrappy so shelter in the form of bogwood or other is recommended. Feeding can be a problem initially but newly-hatched *Artemia* nauplii are eagerly consumed. Once eating, frozen or live bloodworms, glassworms and/or mosquito larvae are taken readily. Spawning accounts are few. Ripe pairs excavate a pit in the gravel and lay about 100 eggs on the bottom. The free-swimming fry are too small for newly-hatched *Artemia* nauplii and growth is slow. The use of RO-processed water will increase the probability of spawning.
- **Comments:** The genus *Biotodoma* was separated from the other Geophagines on the basis of their smaller snouts and the positioning of the mouth. The genus contains one other described, and several undescribed, species. *Biotodoma wavrini*, from Guyana and the Orinoco basin, may be distinguished from *B. cupido* by the positioning of the flank ocellus: on and above the upper lateral line just below the dorsal fin in *B. cupido*, below the lateral line near the body's midline in *B. wavrini*. *B. wavrini* is also slightly more elongated. Care as for *B. cupido*. Usually imported at small size, young *Biotodoma* are typically hollowed-out, grey, nondescript fish. However, with proper care and feeding, they metamorphose into spectacularly contoured adults at about two years of age.

Below: Biotodoma cupido *is a beautiful but demanding cichlid that has rarely been spawned.*

Acarichthys heckelii
Heckel's Threadfinned Acara
- **Distribution:** The Guianas, Amazon drainage
- **Length:** Up to 20cm (8 in)
- **Diet:** Omnivorous
- **Sexing:** Essentially isomorphic
- **Aquarium maintenance and breeding:** Relatively undemanding, *A. heckelii* tolerates a variety of water chemistries and eats anything. They can be belligerent, so careful choice of tankmates and adequate hiding places are essential. In the wild, *A. heckelii* has a most interesting spawning behaviour. Females excavate a series of tunnels in the soft mud of the bottom that lead to a larger 'nuptial chamber', a hollowed-out cave where the eggs will be laid. Once finished, the females actively court males swimming into their territories. Successful pairing results in the attachment of nearly one thousand eggs to the walls and ceilings of the nuptial chamber. These are fanned by the female while the male provides perimeter defence topside. Once hatched and free-swimming, the tunnel system provides the focal point for their

Above: *Heckel's Threadfinned Acara,* Acarichthys heckelii, *excavates tunnels as part of its unusual spawning behaviour.*

foraging and, if threatened, parents and fry retreat into its safety. This unusual spawning mode may be approximated in the aquarium using a large inverted clay flowerpot or large-diameter PVC pipe stood on end. Although difficult to spawn, high temperatures of 29-32°C (85-90°F) and frequent, large water changes seem to initiate spawning. Fry are easy to raise.
- **Comments:** Although closely allied to the Eartheaters, *A. heckelii* lacks a lobed gill arch. The species *geayi*, formerly in *Aequidens*, was placed in *Acarichthys* for a while. More recently, Kullander (1989) has created the genus *Guianacara* for *geayi* and several *geayi*-like fish from the Guianas. *A. heckelii*, with its dramatically produced 'thread-finned' dorsal, is one of the most spectacular Neotropical cichlids in the hobby. Look for it as a contaminant in shipments of *S. sp.* 'jurupari' and/or *G. sp.* 'surinamensis'.

APISTOGRAMMA AND ALLIES

Members of the genus *Apistogramma*, or Apistos for short, are dwarf cichlids ranging in size from 2.5-7.5cm (1-3 in). Like Eartheaters, *Apistogramma* species have a lobed first gill arch with rakers arranged along the inside margin. However, the positioning of their lateral line is closer to the dorsal fin than that in the Geophagines. Despite their affinity with the Eartheaters, Apistos do not sift; rather they pick at small benthic invertebrates, chiefly insect larvae and worms, which form the bulk part of their diet in the wild. In the aquarium, they require live or frozen foods for optimal health. There are nearly 50 described species of *Apistogramma*, and a handful of undescribed forms found all over tropical South America. They live in streams, ponds and oxbow lakes off the main rivers over fine bottoms, often with submerged branches and considerable leaf litter which are used for shelter and the caves in which they spawn. Care for the group is virtually identical and is described below for three commonly available Apistos.

In addition to the Apistos, there are a variety of other dwarf cichlids which are closely allied. These include members of the genera *Microgeophagus* (*Papiliochromis*), *Apistogrammoides*, *Taeniacara*, *Biotoecus*, *Crenicara*, *Dicrossus* and *Mazarunia*. Several of these, particularly the Rams (*Microgeophagus* species) and the Chequerboards (*Crenicara*, *Dicrossus* species), are popular, beautiful aquarium fish the requirements of which are virtually identical to the Apistos.

● **Sexing:** Highly dimorphic. Males nearly twice as large as females, with more highly developed finnage. Their tails are spade-like, and their colours more vibrant. Spawning/brooding females turn bright golden with conspicuous black trim to their ventral fins, a black 'bandit' eyeband, and darkened lateral spot (see photo on page 30)

● **Aquarium maintenance and breeding:** Like most Apistos, these are best handled in planted tanks with peaceful schooling dither fish like tetras, pencilfish, etc. Should be kept clean and warm in temperatures of 26-29°C (78-85°F). Sponge filters are a help in maintaining water quality, as are regular water changes. Best success with most Apistos is by using soft, acid (pH 5-6) water; the use of RO-processed water can be helpful. The addition of peat extract or boiled peat in the filter is recommended to simulate blackwater. Food should be primarily frozen and live, with bloodworms, mosquito larvae, and glassworm larvae alternating. A 'cave' for spawning (an inverted flowerpot with a notched rim, hollowed-out half coconut shell, rock pile) should be provided. In larger tanks of 90-180 litres (20-40 gallons), multiple pairs, or one male and several females, may coexist peacefully. Each female in the harem should be provided with her own cave.

Below: *Male* Apistogramma agassizii *are known by their conspicuous spade-shaped tails.*

Apistogramma agassizii
Agassiz's Apisto
● **Distribution:** Blackwaters of the Peruvian and Brazilian Amazon
● **Length:** Males up to 7.5cm (3 in), females 2.5-3.75cm (1-1.5 in)
● **Diet:** Requires frozen and live foods

• **Comments:** Named after the eminent natural scientist, Louis Agassiz of Harvard, the name is correctly pronounced 'Ag-a-see'-eye'. Several colour variants (local populations) are available in the hobby, including red, gold and blue. These colours are prominently displayed in the dorsal, anal and caudal fins of the male.

Apistogramma cacatuoides
Cockatoo Apisto
• **Distribution:** Streams and oxbow lakes of the Peruvian Amazon
• **Length:** Males to 9cm (3½ in), females half that size
• **Diet:** Requires frozen and live foods
• **Sexing:** Highly dimorphic. Male develops magnificent 'cockatoo crest' in which the first 3-5 spiny rays of the dorsal fin are elongated dramatically. The tail fin of the male is moderately lyrate and may have one or several irregularly coloured spots. Females lack these
• **Aquarium maintenance and breeding:** Maintenance as for *A. agassizii*. Cave-spawning, harem polygynists best maintained in groups of one male to several females. Each female should be provided with her own cave and territorial space. Harems should be maintained in tanks of 140 litres (30 gallons) or larger, but may be

Below: *Males of the Cockatoo Dwarf,* Apistogramma cacatuoides, *develop dorsal fin cockatoo crests.*

spawned as pairs in smaller tanks, with care taken to provide shelter and dither or target fish.
• **Comments:** Like *A. agassizii*, the Cockatoo Apisto is available in several colour morphs which reflect its patchy distribution in the wild. One of these, a red form, is particularly notable for the extensive red spotting that develops in the caudal fin of the male. These are being captively bred and selected for increased colour and spotting in Germany. Wild *A. cacatuoides* are available occasionally in shipments of mixed dwarf cichlids from Peru.

Apistogramma steindachneri
Steindachner's Apisto
• **Distribution:** The Guianas
• **Length:** The largest of the hobby Apistos; males can reach 11cm (4½ in), females 2.5-5cm (1-2 in)
• **Diet:** Omnivorous
• **Sexing:** Males considerably larger than females with slightly lyrate, filamentous tipped tail
• **Aquarium maintenance and breeding:** The most common and easily maintained of the Apistos. While the stringent conditions outlined for *A. agassizii* benefit this fish, Steindachner's Apisto can be maintained and bred on prepared and frozen foods with a little less concern for water chemistry and condition. May be bred in pairs, but may also be polygynous if provided with multiple females.
• **Comments:** A high-bodied, robust Apisto typically mixed with

Above: Steindachner's Apisto, Apistogramma steindachneri, *is among the easiest of the dwarf cichlids to maintain.*

Nannacara anomala in mixed dwarf cichlid shipments from Guyana. They are readily available, inexpensive, hardy and provide a good beginning point for learning about Apisto care and breeding. Named after the early twentieth-century Austrian ichthyologist and student of cichlids, Franz Steindachner.

Microgeophagus ramerizi
Ram, Rameriz's Dwarf, Butterfly Dwarf
● **Distribution:** Streams and ponds of the Rio Orinoco drainage in the Venezuelan and Colombian llanos

● **Length:** Up to 6cm (2½ in)
● **Diet:** Frozen and live foods recommended
● **Sexing:** Moderately dimorphic. 'Cockatoo crest' (elongate first several rays of the spiny dorsal fin) of the male slightly more produced than that of the female. Females slightly smaller and rounder, and when ripe develop a rosy red flush to their ventrum
● **Aquarium maintenance and breeding:** Best maintained as for *A. agassizii.* Unlike most Apistos, *M. ramerizi* is monogamous and spawns in the open, typically on a stone. Fry are often difficult to raise with parents and may require artificial incubation. They are small

Below: The Venezuelan Ram, Microgeophagus ramerizi, *is a beautiful dwarf cichlid.*

and may require liquid fry food, microworms or rotifers for their first feedings. Often high temperatures of 30-31°C (86-88°F) are necessary to elicit spawning.

● **Comments:** First captured and bred by Manuel Ramirez, for whom the fish was named, and Herman Blass in 1947. Initially described in the genus *Apistogramma*, but subsequently removed to the genus *Microgeophagus*. Kullander later proposed the genus name *Papiliochromis* for the Ram, but his position is controversial and, for the time being, *Microgeophagus* would appear to be the correct generic nomen. Aquarists will find additional information in the hobby literature under all three names.

A second species, *M. altispinosa*, the Bolivian Ram has recently appeared in the hobby. A larger, more Geophagine-like fish with a pleasing green/gold cast to the body, *M. altispinosa* grows to a length of 10cm (4 in) and, like its congener, is an open substrate (non-cave) spawner.

Both species are regularly bred in aquariums in Asia, where selective breeding has produced several cultivars of the Ram including golden (xanthistic) and 'veiltail' varieties.

Dicrossus filamentosus
Lyretailed Chequerboard

● **Distribution:** Blackwaters of the Rio Negro, Brazil, and the Rio Orinoco basin of Colombia

● **Length:** Males to 7.5cm (3 in), females to two-thirds that size

● **Diet:** Frozen and live foods

● **Sexing:** Dramatically dimorphic. Males develop lyrate tail fins produced to filamentous streamers on top and bottom and iridescent blue spangled stripes over the central chequerboard band. Females lack the tail streamers and iridescence.

● **Aquarium maintenance and breeding:** Care as for *A. agassizii*. The Lyretailed Chequerboard has been bred infrequently and with considerable difficulty. Should be kept warm, 30-31°C (86-88°F), and in very clean water. As they live in blackwaters with cardinal tetras (*Paracheirodon axelrodi*), the water should be very acid (pH 4-5) and have little, if any, hardness: RO-processed water is helpful. Additionally, peat extract helps to

Below: *The Lyretailed Chequerboard,* Dicrossus filamentosus, *is among the more challenging and beautiful of the dwarf cichlids.*

Above: Crenicara punctulatum *is a larger, higher-bodied chequerboard cichlid.*

duplicate the blackwater environment. The fish often select broad-leafed plants, which are recommended, to support their clutches of 50-150 eggs. Brooding females announce themselves with a dark black longitudinal stripe down the middle of their body. Eggs often fail to hatch, probably due to inappropriate water chemistry. Both fry and adults benefit from feedings of newly-hatched *Artemia* which is an excellent 'first food' for newly-imported, emaciated specimens.

● **Comments:** Described at first as *Crenicara filamentosa* by Ladiges in 1958 from aquarium specimens, Kullander re-described the species in 1978, and has since advised placement in the separate genus *Dicrossus*. Most aquarium references use the older name. There are two other 'chequerboard cichlids' that aquarists may encounter. *Crenicara punctulatum* is a much larger, more robust relative from Peru that may grow to nearly 15cm (6 in). They share the distinctive chequerboard pattern, but they are high-bodied with ovate tails. *Dicrossus maculatus* is a transitional form that hails from the Brazilian Amazon. It shares with *C. punctulatum* the larger size of 10-13cm (4-5 in) and an asymmetrical ovate tail, but is not nearly as high bodied.

Juvenile *D. maculatus* are virtually indistinguishable from *D. filamentosus*. Although the only chequerboard dwarf in the hobby of the 1940s and 1950s, *D. maculatus* has been absent until only recently (1988) when its location was rediscovered by German aquarists and breeding stock returned there. Care of all three species is identical.

PIKE CICHLIDS

Pike cichlids of the genus *Crenicichla* are elongate, torpedo-shaped fish admirably adapted for life as piscivorous ambush predators. They have long snouts with large, teeth-studded protrusible mouths that enable them to grab and swallow smaller fish they vacuum in. Their common name is a tribute to the ultimate ambush fish, the pike (*Esox* species), however, they are related only in the convergence of their anatomy and habits. Pike cichlids are distributed throughout South America and number at least 50 species, including some dwarf forms. Despite their reputation as nasty aquarium residents, in fact the majority of pike cichlids can be kept in the rowdy cichlid community tank with other fish large enough to escape ingestion. Bonded pairs are particularly loyal and exceedingly gentle parents to their offspring, caring for them for 6-12 months in the wild. The tips on maintenance and breeding of the four species reviewed below hold for the majority of pike cichlids.

Crenicichla sp. affin. saxatilis

Spangled Pike

● **Distribution:** Ponds, oxbow lakes, and swamps of the Guianas
● **Length:** Males up to 30cm (12 in), females about two-thirds as long
● **Diet:** Initially live feeder fish, later freeze-dried krill and frozen foods

● **Sexing:** Dimorphic. Males have extensive gold or silver spangling on their flanks and spotting in their dorsal and anal fins, both of which are produced to long filaments. Females have less spangling, have a distinctive white submarginal band in the dorsal fin sometimes with multiple, small ocelli, and develop distended cherry-red bellies when ripe

● **Aquarium maintenance and breeding:** Pike cichlids, in general, are somewhat undemanding in the aquarium. Although predatory in nature, with a little patience they can be converted to a diet of freeze-dried krill, frozen bloodworms, and even some pelleted foods.

● **Comments:** For emaciated, newly-imported specimens, live feeder fish will restore health until dietary conversion is attempted. Adults may be belligerent, so potential spawning partners should be separated until the female ripens and the divided 'pair' show interest in each other. Juveniles may be raised as a group if they are of the same size; smaller individuals will be picked on. Provide sufficient shelter in the form of PVC piping of appropriate diameter and length, at least one tube per fish.

Pike cichlids are cave spawners. In the aquarium acceptable caves include inverted notched flowerpots or stacked driftwood. Several

Below: *This male Spangled Pike,* Crenicichla sp. affin. saxatilis, *has more body spangling than his female consort.*

hundred eggs are attached to the cave via adhesive threads, much like the eggs of West African cichlids (*Pelvicachromis, Nanochromis*) species. Free-swimming fry are huge, growth is rapid and the youngsters are soon eating diced frozen bloodworms and each other. The parents are unusually protective and gentle parents and may continue to provide care for many weeks post-spawning.

● **Comments:** The *saxatilis*-complex contains about 13 or so species which share similar colouration. *Crenicichla saxatilis* is limited in its distribution to the commercially-uncollected areas of Surinam and French Guiana, so its availability in the hobby is doubtful. However, *Cr. albopunctata* from Guyana is nearly identical in appearance and is the 'saxatilis' of the hobby. Several *saxatilis*-complex species, like *Cr. proteus* and *Cr. anthurus*, hail from the Peruvian Amazon, and at least one, *Cr. geayi*, is imported from the Rio Orinoco. All have similar requirements in the aquarium.

Crenicichla regani
Dwarf Pike

● **Distribution:** Brazilian Amazon Blackwater, Rio Tocantins
● **Length:** Males up to 15cm (6 in), females up to 10-12.5cm (4-5 in)
● **Diet:** Small feeder fishes, freeze-dried krill and frozen foods
● **Sexing:** Females alone sport 1-3 large irregular black/white occelli-like splotches in their dorsal fin, and have distended cherry-red bellies

Above: *This female Dwarf Pike,* Crenicichla regani, *is conspicuous for the irregular black and white dorsal fin splotching.*

when ripe. Females about two-thirds the size of males
● **Aquarium maintenance and breeding:** Dwarf Pike cichlids are for aquarists with smaller, planted tanks and may be housed with a variety of characin dithers, as long as these are large enough to escape predation. These should be treated like Apistos and kept in soft, acid blackwater, which is kept warm 26-29°C (80-85°F) and very clean. RO-processed water may be helpful in inducing spawning. They readily accept frozen bloodworms and prosper on them as a staple of their diet. Live dwarf red earthworms may be useful for conditioning. These are cave spawners which have been spawned infrequently. Ripe females develop the distended, cherry-red bellies that signal their intent. Courtship involves head-down, belly-wriggling dancing reminiscent of courtship in the West African Krib, *Pelvicachromis pulcher*.
● **Comments:** There are several species of Dwarf Pikes that grow less than 15cm (6 in). These include the Amazonian species *Cr. notophthalmus*, whose females replace the irregular dorsal splotching of *Cr. regani* with several white-ringed-black ocelli in their dorsal fins, and two from Guyana: *Cr. wallacii* and *Cr. nanus*, which have not yet been in the hobby. Neither of the latter two species is

reported to have dorsal fin ocelli. *Crenicichla heckelii* is the smallest of the dwarfs, reaching only 6cm (2½ in), but this fish has not been imported from the Rio Trombetas into the hobby, nor has *Cr. urosema* from the Rio Tapajos rapids. The beautiful Dwarf Pike, *Cr. compressiceps*, from the Rio Tocantins/Xingu system, has appeared only recently. It grows to only 10cm (4 in) and has been spawned in the aquarium. Dwarf Pike cichlids are definite 'must haves' for Neotropical cichlid enthusiasts.

Crenicichla sp.
Orange Pike, French Fry Pike, Xingu Pike
● **Distribution:** Rio Tocantins/ Xingu system
● **Length:** Member of the *'strigata'*-complex, which reach lengths approaching 46cm (18 in)
● **Diet:** Feeder fish, freeze-dried krill, pelleted prepared foods
● **Sexing:** Females develop distended cherry-red bellies when ripe, and retain white submarginal bands in the dorsal and caudal fins. There is no apparent sexual difference in size
● **Aquarium maintenance and breeding:** Despite their huge adult size, most members of the small-scaled *'strigata'*-complex are peaceful denizens of the large-fish community tank. Bonded pairs are devoted to one another and remain quite peaceful. Spawning has been achieved only infrequently, in a limited number of species. Those spawned, like *Cr. marmorata* have also proved to be cave spawners. Care as for *Crenicichla sp. affin. saxatilis*. These are easily converted to krill or pelleted foods. Members of the *'strigata'*-complex are best acquired in groups of 4-6 juveniles, of near identical size, and raised to adulthood. Be sure to provide enough shelter in the form of PVC tubing.

● **Comments:** Species of the *'strigata'*-complex are notable for their shared juvenile colorational pattern, and for the dramatic colorational metamorphosis they undergo when maturing at around 15cm (6 in) in length. Species of the complex include *Cr. cincta, funebris, johanna, lenticulata, lugubris, marmorata, ornata* and *strigata*, but are typically sold, particularly the young, as 'johanna', 'lugubris' or 'strigata'. Part of the excitement of raising these fishes is seeing just into what the juveniles transform at 15cm (6 in)! A good choice for breeders looking for a challenge, the Orange Pike has yet to be captively spawned.

Below: *This as yet undescribed* Crenicichla *species from the Rio Xingu is often sold as 'Orange Pike' when small.*

Above: *All juvenile pikes of the 'strigata' complex, like this 'Orange Pike', express a shared striped pattern until metamorphosing at 15cm (6 in).*

Crenicichla sedentaria
'Frog-eyed' Pike, 'Hopping' Pike, Sedentary Pike

● **Distribution:** Peruvian Amazon, Ecuador, Colombia
● **Length:** Males to 25cm (10 in). Females half to two-thirds the size
● **Diet:** Small feeder fish, freeze-dried krill, pelleted foods
● **Sexing:** Conspicuously dimorphic. Females with large ocellus in the dorsal fin. Ripe females develop a distended belly and bright red dorsal fin
● **Aquarium maintenance and breeding:** As for the generality of pike cichlids. The 'Frog-eyed' pikes are believed to be cave spawners. PVC piping and/or inverted and notched ceramic flowerpot saucers, large enough to permit access and room to spawn, are recommended.

Have not yet been spawned in the aquarium.
● **Comments:** The genus *Batrachops* was created to hold a subgroup of pike cichlids with a unique arrangement of teeth and thick 'salami-like' bodies. The short snout confers a bug-eyed frog-like appearance to these fish, hence the genus name (batrachians are amphibians). Recently, Kullander has restored species of the genus *Batrachops* to the genus *Crenicichla*, but the former occupants of that genus form a coherent subgrouping of species. The species include: *Cr. cyanotus* (Amazonia), *reticulata* (Amazonia), *semifasciata* (La Plata, Mato Grosso), and *cametana* and *cyclostoma*, both from the Rio Tocantins/Xingu system. *Crenicichla sedentaria* was described by Kullander in 1986, and is apparently *'Batrachops'*-like in many respects. The name *sedentaria* refers to its sedentary habit of sitting on the bottom. All of the 'Hopping Pikes' are moderately to especially rheophilic (rapids-loving), have reduced swim bladders, and spend much of their time hugging the bottom and hopping, though all can swim. Those from the Tocantins/Xingu system prosper from the use of submersible powerheads.

Below: *This female* Crenicichla sedentaria *is a commonly-available 'Hopping' Pike from the* Batrachops *complex.*

THE CICHLASOMINES

Cichlasomines of the genus *'Cichlasoma'* have radiated dramatically in Central America from South American ancestors that crossed the isthmus of Panama several million years ago: there are now over 100 described species from Texas down through Panama. However, the number of Cichlasomines is decidedly more modest in South America proper, where the Acaras and their descendants have competed effectively for various ecological niches. Nevertheless, fish like the Angelfish, Discus and Uaru represent a few of the more creative evolutionary derivatives of the Cichlasomine lineage in South America.

Originally, fish having four or more hard spiny rays in their anal fins were relegated to the genus *Cichlasoma*. As was true for the three-spined Acaras, this was a rather heterogeneous collection of cichlids. It included, among others, a rather generalised, Acara-like fish named *Cichlasoma bimaculatum*, — the Guyanese 'Black Port' of the aquarium trade. Kullander realised that there were more similarities between the Port Acara *Aequidens portalegrensis* and the Black Acara than the simple difference in anal fin ray counts that divided them. In 1983, he lumped the two species and a number of other 'Port'-like species together in a common genus. The name of that genus was none other than *Cichlasoma*, since *C. bimaculatum* had been the author Swainson's original type-specimen for the genus in 1839!

This redefinition of the genus *Cichlasoma* made nomenclatural orphans of the rest of the Cichlasomine cichlids. For a while, the name *Heros*, next in historical line, was used in place of *Cichlasoma* for these orphaned cichlids. However, in 1986, Kullander redefined and restricted that name to the Severum-like Cichlasomines. In addition, he resurrected old names for many of the sub-groupings. These usages are reflected here. Unfortunately, many of the Cichlasomine cichlids, particularly the Mesoamerican species, have yet to be completely re-evaluated and, for the time being, should be referred to as *'Cichlasoma'* species, with quotation marks around the generic name. Note, however, that the aquarium literature has yet to follow suit, and that most of these fish can be accessed only with their older *Cichlasoma* nomen.

Cichlasoma portalegrense
Port Cichlid

- **Distribution:** La Plata Basin, Argentina, southern Brazil
- **Length:** Up to 15-20cm (6-8 in)
- **Diet:** Omnivorous
- **Sexing:** Essentially isomorphic Ripe females more rotund than males

- **Aquarium maintenance and breeding:** One of the easiest of the biparental substrate-spawning cichlids, hardy, 'industrial-strength' fish whose maintenance and dietary requirements are quite basic and easily met. 'Ports' are exemplary parents, and may be maintained at lower temperatures of 20-24°C (68-75°F) with brief drops to 15°C (60°F) because of their subtropical distribution. *Cichlasoma portalegrense* was historically one of the earliest cichlids in the hobby, making their appearance in Europe in the early 1900s.

- **Comments:** Formerly known as *Aequidens portalegrensis*, under which name its care has been described for decades in the hobby literature. *Cichlasoma bimaculatum*, the Black Port, is a frequent import from Guyana, and several of the other *Cichlasoma* species *sensu stricto* (eg *C. taenia, amazonarum, paranense, araguaiense* and *dimerus*) are occasional accidental contaminants, and are usually sold as 'Ports'. Occasional 'movable

Above: *The Port Cichlid*, Cichlasoma portalegrense, *was one of the first cichlids to be exported from South America, in the early 1900s.*

platform' spawners, choosing leaves when available.

'Cichlasoma' festae
Red Terror
● **Distribution:** Pacific slope of Ecuador
● **Length:** Males up to 46cm (18 in), females two-thirds that size
● **Diet:** Omnivorous
● **Sexing:** Conspicuously dimorphic. Males develop green iridescence on their flanks and in older specimens, 'craggy cheeks' (corrugated preopercle) and thick lips appear, eliciting the common native name for this species:

'vieja' or 'old woman'. Females remain more rotund, smaller, and maintain the alternating bright red and black barring of juvenile fish
● **Aquarium maintenance and breeding:** These fish have rightly earned their common name 'Red Terror': they are extremely belligerent. Adults must be housed in large tanks with rowdy cichlid tankmates. Even so, 'pairs' may routinely liquidate each other and the fish they are housed with. For larger specimens, the 'partial-divider' method is the best propagation strategy. Alternatively, a group of juveniles may be raised to sexual maturity and allowed to form

Below: *This courting female Red Terror,* 'Cichlasoma' festae, *is absolutely incandescent.*

compatible pairs. They will spawn at 10-15cm (4-6 in) despite their huge maximal size. These are typical biparental substrate spawners that lay several hundreds of eggs. Nothing compares with the incandescent colouration of a brood-tending female *'C.' festae* — the vertical bars turn bright orange-red alternating with black. For all their aggression, these are gentle and excellent parents and a joy to watch. They should be handled like the majority of large Cichlasomines in terms of water maintenance and feeding.

● **Comments:** The Cichlasomines from west of the Andes in northwestern South America are quite unlike those to the east, and more closely resemble their relatives from Central America. *'Cichlasoma' festae* clearly resembles the drabber Central American species, *'C.' uropthalmus*, both in coloration and habit and is often confused with it in the hobby, usually to the advantage of the seller. Two other generalised Cichlasomines from northwestern South America include *'C.' atromaculatum*, from the Rio Atrato in Colombia (see page 27 for photograph) and the extremely rare *'C.' ornatum* (Colombia, Ecuador). Neither fish is particularly common in the hobby.

Caquetaia spectabilis
Spectabile, False Basketmouth
● **Distribution:** Amazonia, Rio Tocantins/Rio Xingu system, Guyana
● **Length:** Up to 25-30cm (10-12 in) in the aquarium
● **Diet:** Gape-and-suck ambush predators in the wild, omnivorous in the aquarium
● **Sexing:** Essentially isomorphic
● **Aquarium maintenance and breeding:** Beautiful but somewhat delicate fish. Particularly sensitive to water quality, they respond to lax maintenance by neuromast 'pitting'. Despite their apparent piscivory, they will prosper on a diet of pelleted prepared foods, freeze-dried krill and a variety of frozen foods. Earthworms may be useful in conditioning adults. Spawns are huge, numbering from 500-1000 depending on the size of the adults. Pairs are excellent parents.
● **Comments:** Despite its widespread distribution, *C. spectabilis* is rather uncommon in the hobby and is rarely imported from the wild, which is unfortunate as it is one of the most beautiful of the South American Cichlasomines. There are two other species in the

Below: *The Spectabile,* Caquetaia spectabilis, *is a rare and somewhat delicate Cichlasomine.*

genus *Caquetaia. Caquetaia
kraussii* hails from Venezuela and
Colombia (Rio San Juan, Rio Atrato
basins), and is a less colourful,
bronze-brown version of *C.
spectabilis*. The third species, *C.
myersi*, is also from the Colombian
tributaries of the Amazon and
Orinoco basins, but has been
absent from the hobby. They share
similar requirements in the
aquarium, but only *C. spectabilis*
and *C. kraussii* have been spawned
and are occasionally available. The
genus name *Caquetaia* derives
from the Rio Caquetà in Colombia,
the site of original capture of the
type specimen for the genus. These
fish were once placed in the genus
Petenia, along with *P. splendida*
from Central America, an ecological
analogue with a similarly protrusible
mouth and piscivorous habit.

Hypselecara temporalis
Chocolate Cichlid
● **Distribution:** Peruvian and
Brazilian Amazon
● **Length:** Up to 25cm (10 in)
● **Diet:** Omnivorous
● **Sexing:** Males with nuchal hood
on forehead
● **Aquarium maintenance and
breeding:** Chocolate cichlids can
be rather belligerent, particularly
with their own. The difficulty in
raising and spawning them,
therefore, lies in establishing a
compatible pair. While adults of
both sexes can be paired by the

Above: *This female Chocolate
Cichlid,* Hypselecara temporalis,
*shown here tending eggs, lacks the
nuchal hump of her mate.*

'blind date' method, a better
solution is to raise a group of
juveniles up to sexual maturity.
Other, similarly aggressive, target
fish help to cement the pair bond
and dispel intra-pair aggression. In
addition, this species is
reproductively-precocious and will
spawn at 10-13cm (4.5 in), despite
the much larger maximum adult
size. Some attention should be paid
to water quality and chemistry, but
most foods are eaten ravenously
and growth is rapid. Egg-eating is
typical of young pairs but
disappears with experience. Raising
the fry is straightforward, and there
is a ready market for the hundreds
of juveniles each spawning can
generate.
● **Comments:** Kullander created
the genus *Hypselecara*, meaning
'High Acara', to commemorate the
egg-shaped body of this fish. The
chocolate cichlid has been known
under several names both in
science and in the hobby. These
include *(Cichlasoma) goeldii,
crassa, hellabrunni,
coryphaenoides, arnoldi* and *Chuco
axelrodi*. With the exception of
coryphaenoides, now also in
Hypselecara, the rest have been
synonymised with either *H.
temporalis* or *H. coryphaenoides*.

Typically, these have been geographic colourational variants elevated to species status. *H. coryphaenoides* is found in the Negro, Orinoco and Trombetas drainages and is readily distinguished from *H. temporalis*. The juvenile fish present a sharply-chiselled head profile, often with a light-coloured blaze down the ridge, and have a midlateral spot or band that extends vertically to the upper lateral line.

H. coryphaenoides is a species occasionally imported, but *H. temporalis* is the usual chocolate cichlid of the hobby, being bred commercially in Florida and Asia.

Heros severus
Severum
- **Distribution:** Guyana, Amazonia, Colombia, Venezuela
- **Length:** Up to 25-30cm (10-12 in)
- **Diet:** Omnivore, vegetable material recommended
- **Sexing:** Essentially isomorphic. Males more elongate than females. In some populations, males with more extensive vermiform spotting on flanks
- **Aquarium maintenance and breeding:** Care as for the chocolate cichlid, *Hypselecara temporalis*. Water should be kept clean and warm around 26-29°C (78-84°F), and the fish fed a variety of plant material (eg romaine, lettuce, spinach) in addition to the regular prepared pelleted, frozen and live foods (earthworms). Severum can be belligerent, especially as adults. Compatible pairing may be difficult with the 'blind date' method, even if the adults are kept divided for some time.

The presence of target fish may increase the chances of success. These are typical biparental substrate spawners which spawn precociously at a size of 10cm (4 in) at between 12-18 months of age. There is always a good market for young of this species, particularly the cultivar gold (xanthistic) variety.
- **Comments:** Kullander, in 1986, restricted usage of the genus name *Heros* to the species *severus* and elevated at least one colorational variant from Peru, to species status (eg *H. appendiculatus*). It is likely that other geographic variants will receive species status in the future. In addition to wild specimens, *H. severus*, both as green and xanthistic gold forms, are bred in Asia and Florida and are commonly available in the trade.

Below: *The Green Severum,* Heros *severus, is commonly available in the trade.*

Mesonauta festivus
Festivum, Flag Cichlid
- **Distribution:** Guianas, Orinoco and Amazon basins
- **Length:** Up to 15-20cm (6-8 in)
- **Diet:** Omnivorous
- **Sexing:** Essentially isomorphic. Males with longer 'nose'
- **Aquarium maintenance and breeding:** Festivum are typically found in association with Angelfish, *Pterophyllum scalare*, in the wild, and should be maintained like wild angelfish. Soft, acid water is helpful, but not essential. Because of their shyness, they should be kept in a planted tank with some shelter, perhaps bogwood. They neither dig nor eat plants. They are relatively peaceful and can be kept with other less-aggressive cichlids, like Eartheaters or Acaras, or dithers large enough to escape being eaten. Spawns are of modest size (300-500 eggs) and pairs are reliable parents once induced to spawn, which can be difficult. Males often have a larger, more exaggerated 'Roman nose'. A nice alternative to the larger and more aggressive South American Cichlasomines.

Above: *The Flag Cichlid,* Mesonauta festivus, *is found with Angelfish in the wild.*

- **Comments:** Kullander, in 1986, resurrected the genus *Mesonauta* for the species *festivuus*. Noting the extreme geographic variability of Festivum, Kullander, in 1991, split the species into five. These include *insignis* from the upper Rio Negro and Rio Orinoco, *acora* in the Tocantins and Xingu drainages, *egregius* in the Colombian Orinoco basin, *mirificus* in the Peruvian Amazon, and *festivus*, now restricted to the Paraguay and Bolivian Amazon basins, the Rio Jamari and lower Rio Tapajos. They differ in body proportions and markings, and there are several subtle differences in spine count, etc. At the time of this publication, Guyanan and Brazilian Amazonian specimens have not yet been studied, but it is likely that Kullander will eventually treat these as discrete species. Most of the commercially available festivum are bred in Asia or Florida, with the remainder wild fish, usually from Guyana.

Hoplarchus psitticus
Parrot Cichlid
- **Distribution:** Amazon and Orinoco basins
- **Length:** Older adults to at least 40-46cm (16-18 in)
- **Diet:** Omnivorous, some vegetable material recommended
- **Sexing:** Essentially isomorphic. Males slightly more elongate than females with a slight nuchal crest as they grow large.
- **Aquarium maintenance and breeding:** Care as for the generality of blackwater species. Soft, acid water kept exceptionally clean and warm, 26-30°C (78-86°F), is a must: these fish are particularly susceptible to neuromast erosion (Head Hole). Adults can be aggressive towards each other and pairing can be a problem. They have been spawned only a few times in the aquarium. Eggs are very few in number (50-100), are very large and are typically placed on a vertical surface. Because the parents are prone to eating their spawns, what fry have been raised have been done so artificially. The fish are quite shy and easily distracted from spawning by the presence of other fish, or by placement of their tank in a public venue. It seems also that reproductive maturity is reached only after considerable age at about 3-5 years.

- **Comments:** Available only very infrequently, typically as accidental contaminants with chocolate cichlids *Hoplarchus psitticus* is apparently rare in the wild. The Amazonian specimens are the more colourful than their Colombian counterparts. Inclusion of carotene-containing foods (eg krill) will help bring out the bright red highlights on this otherwise iridescent green fish. They also appreciate some 'greens' in their diet.

'Cichlasoma' facetum
Chanchito, Zebra Cichlid
- **Distribution:** Southeastern Brazil, Argentina, Uruguay, La Plata drainage
- **Length:** Up to 15-20cm (6-8 in)
- **Diet:** Omnivorous
- **Sexing:** Essentially isomorphic
- **Aquarium maintenance and breeding:** Despite its smaller size, the Chanchito exhibits all of the behavioural quirks of the larger Cichlasomines: it is aggressive and it digs. However, it makes a reasonable addition to a not overly rowdy community of cichlids. Chanchito means 'little pig' and aptly summarises the feeding behaviour of this fish: they are aggressively and enthusiastically

Below: *The Parrot Cichlid,* Hoplarchus psitticus, *grows to around 40cm (16 in).*

Above: *Despite its smaller size, the Chanchito, 'Cichlasoma' facetum, can be as aggressive as its larger relatives.*

omnivorous. Breeding is as for the generality of biparental substrate-spawning cichlids. They are easily induced to spawn and are generally excellent parents. Hailing from subtropical South America, these fish can take lower temperatures of 20-24°C (68-75°F), even briefly down to around 13°C (55°F). This is one reason why the Chanchito was historically the first successfully-kept aquarium cichlids in the mid-1890s.

● **Comments:** There are more 'Chanchito-like' fishes which may well be distinct species: *'Cichlasoma' autochthon* and *'C.' oblongum*, somewhat smaller than *'C.' facetum* at 13-15cm (5-6 in) and with vertical barring of variable colour and intensity. This fish is not propagated commercially, consequently we must rely on infrequent accidental importations from Argentina.

Pterophyllum altum
Altum Angelfish
● **Distribution:** Rio Orinoco, Rio Negro
● **Length:** Up to 20-25cm (8-10 in)
● **Diet:** Omnivorous. Frozen and live foods recommended
● **Sexing:** Essentially isomorphic
● **Aquarium maintenance and breeding:** *Pterophyllum altum*, the deep-bodied relative of the more common angelfish, *Pterophyllum scalare*, is occasionally imported from Colombia and may be distinguished from *Pt. scalare* by its

higher body, peculiar long, curved snout, and somewhat indistinct brown vertical barring. Angelfish, in general, are found both in whitewater or turbid blackwater, usually in lakes or other lentic habitats, typically with densely vegetated shores punctuated with fallen submerged branches. Whereas *Pt. scalare* yielded to the breeder's skill in the early 1900s and is today the staple of the aquarium hobby, having been captively inbred with strains selected and 'fixed' for colourational and finnage mutations over the past 80 years, *Pt. altum* has to date resisted all attempts. For the serious hobbyist wishing a challenge, the Altum Angelfish is it! Generous feedings of live foods and soft, RO-processed blackwater kept warm, and very clean water will undoubtably be required. *Pterophyllum altum* are best maintained as a colony in their own separate, large tank, chosen to maximise front-to-back depth and with adequate bogwood shelter and planting.

Below: *The Altum Angelfish,* Pterophyllum altum, *is a delicate though beautiful species.*

aquarium and they tolerate a much wider range of water chemistries and condtions. Moreover, they are usually kept in the sparest of conditions: typically, bare tanks with a slate or spawning cone to hold their eggs. Wild Discus should be treated as the blackwater species they are and kept in soft, acid, warm water, around 27-30°C (80-86°F), with added peat extract and kept very clean with effective biological filtration and regular water changes. They prefer tanks with larger bottom areas and shelter in the form of a tangle of bogwood and/or plants to offset their shyness. Dither fish, principally tetras, make them feel more secure. Peaceful cichlids, including *Apistogramma* species, may also be kept with them. One tankmate that should be excluded is wild Angelfish of any species which often carry diseases fatal to Discus.

Discus prosper on a diet of frozen insect larvae (bloodworms, mosquito, glassworms) and live foods. Spawning is another matter, however. Wild discus are difficult to induce to spawn in the aquarium. A rich diet and RO-processed water kept warm and clean are helpful but not necessarily sufficient. Nevertheless, spawnings of wild fish still occur, and many of the various coloured cultivated strains have been engineered by occasional and calculated outcrossing with selected exceptional wild fish. When they do spawn, wild pairs often make exceptional parents and support the initial growth of their fry via contact feeding — a behaviour pattern in which the young fish 'nip-off' nutritional mucus from both parents' sides. Growth, thereafter, is rapid on a diet of newly-hatched *Artemia* nauplii.

- **Comments:** The several colour varieties of *S. aequifasciatus*, brown, blue and green, were given subspecific status by several

- **Comments:** There is one other recognised species of Angelfish: *Pt. dumerilii* from the Amazon. Kullander (1986) believes *dumerilii* to be a junior synonym of *scalare* and recognises the species *Pt. leopoldi* as valid. It is conspicuous for the dark black spot located at the dorsal tip of the fourth vertical bar, just near the site of dorsal fin insertion. It is also rather 'dumpy' (low-bodied), relative to both other species.

Symphysodon aequifasciatus

Discus, Pompadour
- **Distribution:** Blackwater Amazonia, Tocantins/Xingu system
- **Length:** Up to 20-25cm (8-10 in)
- **Diet:** Frozen and live foods recommended
- **Sexing:** Essentially isomorphic. In some populations (eg blue), males with more pronounced vermiform markings on flanks.
- **Aquarium maintenance and breeding:** Like the Angelfish, *Pterophyllum scalare*, Discus have become readily available in the hobby as tank-raised cultivars whose requirements are best learned from speciality books dedicated to these fish. Modern, selected Discus are very different fish than their wild progenitors. For one, they are much more easily induced to propagate in the

Above: *The Discus,* Symphysodon aequifasciatus, *is the king of aquarium fish.*

ichthyologists (eg *a. axelrodi, a. haraldi, a. aequifasciatus*), but the modern view is that they are one species with several geographic variants, all of which can interbreed. A second species, *S. discus,* better known as the True or Heckel Discus, seems distinct. The Heckel Discus, found also in the Amazon, is characterised by somewhat larger and higher body, flanks completely vermiculated by iridescent stripes, and three distinct black vertical bars, the thickest and most distinct through the centre of the fish. These also come in several colour varieties. *Symphysodon aequifasciatus* and *S. discus* have occasionally been cross-bred and the resultant offspring have proven fertile, calling into question the biological distinctiveness of the two species. However, wild hybrids have not been found. Care of *S.discus* is identical to that of *S. aequifasciatus;* however they seem to enjoy higher temperatures of up to 32°C (90°F).

Uaru amphiacanthoides
Uaru, Triangle Cichlid
- **Distribution:** Amazon drainage
- **Length:** At least 30cm (12 in)
- **Diet:** Omnivorous, vegetable material recommended
- **Sexing:** Essentially isomorphic
- **Aquarium maintenance and breeding:** Often associated with Discus that form colonies around submerged trees or branches. Care is that of Discus. They require warm 26-30°C (78-86°F), very clean, soft, acid water. Because these are rather dramatic herbivores, vegetable matter in the form of lettuce, or spinach should be provided regularly. Uaru are biparental substrate spawners and have been spawned regularly in the aquarium, but are notorious egg eaters.

Often, two females will 'pair up' and spawn regularly, alternating egg-laying roles. Artificial incubation of eggs may be necessary to prove the identity and fertility of the 'pair'. When and if they get it right, Uaru are excellent parents, providing the first nutrition for their brood in the form of nutritional mucus that the fry

Above: *The Triangle Cichlid*, Uaru amphiacanthoides, *requires lots of vegetable material in its diet.*

graze from the surface of the flanks: Uaru, like Discus, contact feed. Although sometimes called 'Poor Man's Discus', Uaru are highly desirable and sought-after cichlids in their own right, and there is a steady market for tank-reared young. Pairs are best had by raising a group of juveniles to sexual maturity. One interesting aspect of their development is the colorational metamorphosis that youngsters

Below: *Juvenile Uaru go through a distinctive colorational metamorphosis as they mature.*

undergo as they mature. Juvenile Uaru are dark brown with regularly-spaced white blotches scattered on their flanks. As they mature, these spots enlarge to blotches as the base colour lightens to café-au-lait brown. The adult fish retain this lighter brown colour and develop a black 'eyebrow' mark along with the large black flank triangle from which the common name is taken.

● **Comments:** *Uaru* species have been reported from the Tocantins/Xingu system and from Venezuela. The latter was described as *U. fernandezyepezi* by Stawikowski in 1989. One undescribed species has a rectangular blotch replacing the triangle on the flanks, and lacks the black 'eyebrow' marking.

Cichla sp. affin. ocellaris

Lukanani, tucanaré, Peacock Bass

● **Distribution:** Amazon basin, Orinoco basin, Tocantins/Xingu system, Guianas

● **Length:** Up to 61cm (24 in) in the wild

● **Diet:** Piscivore. Will convert to krill and frozen fish in the aquarium

● **Sexing:** Essentially isomorphic. Large 'bull' males reportedly develop a pronounced nuchal hump just prior to spawning

● **Aquarium maintenance and breeding:** Because of their large size, *Cichla* species have been kept only by the most devoted of specialists. As 'cute' juveniles they seem to accept only live fishes or live Tubificid worms. Adults may, with some trouble, be moved on to a diet of freeze-dried krill, and frozen prawns. They are messy eaters and the water must be kept clean and moving. Reports by Lowe-McConnell (1969) suggest that in the wild *Cichla* species pairs dig several circular nests on the bottom and lay between 6000 and 10000 eggs in one of them. They are devoted parents and guard their young until they reach a length of 4cm (1½ in).

The large ocellated tail spot provides an orienting signal for the fry. Additionally, this eye-like spot is believed to confuse piranhas who make much of their living tearing pieces of fins from fish. These cichlids are recommended only for aquarists with the largest of tanks and much dedication. Perhaps this species is best left in South America where they belong.

● **Comments:** Historically, two species of *Cichla* were recognised: *C. ocellaris* and *C. temensis*. Recently, Kullander (1986) has proposed splitting the species *ocellaris*, which exhibits dramatic regional polymorphism, into a series of discrete species including

Below: *The tucanaré,* Cichla sp. affin. ocellaris, *is a prized game fish throughout South America.*

monoculus (Peru, Amazon), intermedia, orinocensis (Venezuela) and, of course, ocellaris (Guianas). In the absence of reliable information as to the origin of export, the name Cichla sp. affin. ocellaris is best applied to this fish. Cichla species are beloved food and game fishes in their native South America, and have been introduced into Central America for the same purposes.

Astronotus ocellatus
Oscar
● **Distribution:** Amazon and Orinoco basins, French Guiana, Northern Paraguay. Some of these may be distinct species other than ocellatus
● **Length:** Up to 30-36cm (12-14 in)
● **Diet:** Omnivorous
● **Sexing:** Essentially isomorphic
● **Aquarium maintenance and breeding:** Unfortunately, young Oscars are often among the first fish new hobbyists acquire for their 45-litre (10-gallon) community tank

Above: Cichla temensis is notable for its body spotting.

because, at 2.5-5cm (1-2 in), their large eyes and social demeanour make them 'so cute': too bad for the Oscars because they are potentially large fish with special requirements. Usually they perish, and the ornamental trade and breeders count on that eventuality to ensure repeated replacement. If treated well, an Oscar can live for ten years or more, and will outgrow its 10-gallon tank in a few months. Young Oscars should be raised as a group (it is a gamble to attempt to 'blind date' large individuals) in as large a tank as possible. The water should be heavily filtered, as these are messy eaters, and partial regular water changes must be done. Lax attention to water quality inevitably results in neuromast erosion (pitting) which is usually not reversible. The bigger the tank, the less of a potential problem this is. Oscars will readily take pelleted prepared foods and freeze-dried krill and do not need living feeder fish to prosper. The Oscar is prone to rearranging its tank, so care should be taken in the placement of rock piles that could be undermined, fall and ultimately damage the fish or crack the tank. Heaters should be placed and fastened with an eye to destruction or jettisoning. As they grow and reach reproductive maturity in 1-1½ years, they begin pairing off. Housing a pair or group in a 570-litre (125-gallon) tank with compatible target cichlids is not overkill. Once bonded, they may stay together for as long as ten years, spawning all the while. They lay 1000 or so opaque, white eggs (good eggs look like bad, fungused eggs) on a previously-cleaned substrate and both parents participate in their care and defence. The fry grow rapidly and are eminently saleable at a size of

Below: *The Oscar,* Astronotus ocellatus, *is a large 'personality' fish which endears itself to its owners.*

2.5cm (1 in) for the reasons outlined above. The Oscar is a 'personality' fish and soon learns to interact with its owner as he/she becomes associated with food.

● **Comments:** Kullander (1986) suggests that there are several distinct species of *Astronotus* as yet to be described as well as a second recognised species: *A. crassipinnis*, from Peru. In general, these differ in fin and ray counts, and in colourational pattern (ie absence, presence and extent of orange ocellations). The Oscar is another cultivated ornamental fish which has been selected for mutations in colouration and finnage. The Tiger and Red Oscars are two such examples where the extent of red colouration has been improved on over wild fish. More recently, an albino strain has been developed and fixed. In addition, a 'veil-tailed' mutation has been introduced into most of the colour types. These are typically bred in Asia and Florida and are the primary source of hobby Oscars. These latter are particularly tolerant of a wide range

of water chemistries. Wild specimens, which are considerably less colourful, are only occasionally imported, and are considerably less forgiving.

Acaronia nassa
Basketmouth
- **Distribution:** Guianas, Amazon Basin
- **Length:** Up to 20cm (8 in)
- **Diet:** Piscivorous, can be shifted to freeze-dried krill or frozen foods
- **Sexing:** Essentially isomorphic. ripe females heavier, less elongate than males
- **Aquarium maintenance and breeding:** The Basketmouth makes its living as a gape-and-suck ambush piscivore in the wild, hence its common name. The protrusible mouth can be extended in a flash, and the vacuum created is sufficient to suck in swimming prey quickly and decisively. In the aquarium, Basketmouths will learn to take large freeze-dried krill, earthworms and even some pelleted foods, but living feeder-fish may be useful for conditioning potential breeders. They are undemanding about water chemistry, but attention should be paid to water quality. Adults may be somewhat aggressive with each

other, but the Basketmouth can be successfully kept in mixed cichlid communities. They do require shelter into which they can retreat: flowerpots or PVC piping. Captive spawning has been reported only once (Leibel, 1985), and in that instance the fish behaved as a typical biparental substrate spawner. Spawning was effected by cyclical temperature increases to 33°F (92°F) and decreases to 22°C (72°F) every few days over the course of several weeks. The fry grew quickly on an initial diet of newly-hatched *Artemia,* and were switched to diced frozen bloodworms within two weeks. Basketmouths are rarely seen in the hobby, entering typically as single accidental contaminants. They are, however, quite common in the wild.
- **Comments:** A second species, *Acaronia vultuosa,* has recently been described from the Orinoco drainage, which differs principally in the spotting pattern of the head.

Chaetobranchus flavescens
Combtail Basketmouth

Below: *The Basketmouth,* Acaronia nassa, *is a gape-and-suck ambush predator.*

- **Distribution:** Amazon drainage
- **Length:** Up to 25-30cm (10-12 in)
- **Diet:** Freeze-dried krill and a variety of frozen and prepared foods
- **Sexing:** Essentially isomorphic. Both sexes have dramatically produced fin filaments on the dorsal and caudal fins. Males slightly more elongate
- **Aquarium maintenance and breeding:** Spawning habits unknown as it has yet to be captively propagated. Like the 'true' Basketmouth; these have trapdoor mouths; however, the generic name refers to their long, thin gill rakers which suggest a life of plankton-sifting for this fish. Luckily, *Ch. flavescens* readily take freeze-dried krill and a variety of offered frozen foods in the aquarium. Care as for *Acaronia nassa*. Adults of both sexes develop long filamentous streamers on the edges of their tail fins which are echoed in the multi-filamented dorsal streamers, hence the common name. A unique and beautiful fish.
- **Comments:** A second species in the genus, *Ch. semifasciatus*, is

Above: *The Combtail Basketmouth,* Chaetobranchus flavescens, *has never been bred in captivity.*

occasionally imported from the Brazilian Amazon and may be distinguished from *Ch. flavescens* by the presence of a prominent ocellus on the caudal peduncle and several partial vertical bars on the flanks. A second closely-related genus, *Chaetobranchopsis*, with three valid species — *australis* (Argentina, Paraguay), *bitaeniata* (Amazon) and *orbicularis* (Amazon) — is distinguished on the basis of fin ray counts, but, like *Chaetobranchus*, members have the setiform gill rakers characteristic of plankton feeders. These fish apparently do make their living straining small zooplankton and, thus, do poorly in the aquarium. Occasionally, exceptional individuals can be converted to powdered krill or even larger types of food, but typically, *Chaetobranchopsis* species in captivity suffer a long, protracted starvation death.

Index to fishes

Page numbers in **bold** indicate major references, including accompanying photographs. Page numbers in *italics* indicate captions to other illustrations. Less important text entries are shown in normal type.